Everything You Need to Know Now About Gold and Silver

EVERYTHING YOU NEED TO KNOW NOW ABOUT GOLD AND SILVER

EDITED BY LOUIS E. CARABINI

ARLINGTON HOUSE·PUBLISHERS
NEW ROCHELLE, N.Y.

Manufactured in the United States of America

Library of Congress Cataloging in Publication Data

Carabini, Louis E. comp.
Everything you need to know now about gold and silver.

Reprint of interviews from the Gold & silver newsletter.
Bibliography: p.
1. Precious metals—Addresses, essays, lectures. 2. Gold—Addresses, essays, lectures. 3. Silver—Addresses, essays, lectures. I. Gold & silver newsletter. II. Title.
HG261.C37 332.4'22 74-17120
ISBN 0-87000-281-3

Contents

Introduction

Louis E. Carabini, Editor

Remember that silver dime you used to buy a cup of coffee with? It's worth more than 35¢ today. And the U.S. $20 gold piece—also known as a Double Eagle—is worth more than $250.

Gold and silver prices have tripled during the past two years, as increasing numbers of people are turning to these precious metals. I can attest to this surging interest from my own experience as president of the Pacific Coast Coin Exchange. We started eight years ago as a numismatic store with a small side business of supplying silver coins to investors. Now it's a $270 million-a-year business serving people in four countries.

Since both precious metals have been with us for 6,000 years, it's natural to ask: Why the tremendous interest? And why have prices increased so dramatically?

The primary reason is inflation. Whether shopping for food, buying gas or borrowing money, almost everyone feels the pinch of inflation. And everyone with savings accounts, bonds, annuities and other fixed-currency assets are seeing their savings wither away. So more and more people are losing confidence in the ability of governments to check inflation. When that happens, they seek the security of gold and silver. This is why:

First, gold and silver are desired around the world for their great beauty and utility. Jewelry, art and religious objects, and the most beautiful coins are made of gold and silver.

Second, gold and silver are rare. The supply of these precious metals cannot be arbitrarily inflated like paper money; they must be laboriously mined and refined.

Third, gold and silver are durable. Gold and silver coins recovered from Spanish treasure galleons look almost new despite 300 years in salt water. On the other hand, paper money, rare stamps and books can be destroyed by mildew, water or fire; a Ming vase will shatter, a diamond will evaporate in a fire.

For these reasons, gold and silver serve as a *store of value*. Thus, since long before the Bible, people have yearned for gold and silver in emergencies: to buy food during famines; water during draughts; safe passage during wars; and financial protection against inflation, depression and monetary collapse.

Another reason why so many people are buying gold and silver today is that both precious metals are very much in demand for a multitude of commercial applications. Moreover, the demand exceeds mine production, and despite record high gold and silver prices, production is actually falling. So the supply situation is tight.

But, while the interest in gold and silver became widespread, there was no single book a person could turn to for authoritative information covering the entire field. Thus, the idea for this book.

Arlington House was aware that the Pacific Coast Coin Exchange has been publishing the *Gold & Silver Newsletter* for several years. One of the *Newsletter's* most popular features is in-depth interviews with the world's leading experts. Tens of thousands of requests for copies have been received, and more than a few newspapers and national magazines have reprinted selections. So it seemed only natural to gather these interviews in one convenient book, which would explain in nontechnical English virtually everything you need to know about gold, silver and inflation.

The gentlemen you will meet are:

DR. MURRAY N. ROTHBARD, author of *America's Great Depression, What Has Government Done to Our Money?*

and other economics books. He talks about inflation—what causes it and why it goes on year after year. He also analyzes the consequences of inflation, including price controls, shortages, balance of payments deficits, trade wars and depression.

DANA L. THOMAS, best-selling author of *The Plungers & the Peacocks* and *The Money Crowd.* He provides us with historical perspective, telling us about inflation in ancient Rome, France and Germany. He touches on inflation in Argentina and Chile, then talks about the plight of the dollar today.

THOMAS J. HOLT, an investment advisor widely respected for the accuracy of his economic and market forecasts. Mr. Holt explains why he expects to see more inflation followed by a depression. As one who has lived through runaway inflation, Mr. Holt speaks with special authority on gold and silver. He projects the probable course of gold and silver prices for the rest of this decade. His $5 per ounce prediction for silver has already been exceeded.

DR. FRANZ PICK, perhaps the world's foremost authority on currency and gold. He is the author of numerous books and publisher of *Pick's World Currency Report.* Dr. Pick talks about the two most recent dollar devaluations and a new currency to replace the dollar. He discusses runaway inflations in America, France, China, Brazil and six other countries. His prediction that there would be a stoppage in the flow of Arab oil to America came true. Most important, Dr. Pick explains hedging with gold and silver, and recommends four immediate purchases.

HANS WEBER, managing director of the Foreign Commerce Bank, Zurich, Switzerland. He explains why Switzerland is the banking capital of the world; discusses banking privacy; and concludes by recommending a model portfolio for conservative investors.

CHARLES R. STAHL, one of the leading experts on silver and publisher of *Green's Commodity Market Comments.* He reviews the four main areas of silver consumption, and eight little-known, yet important, uses for silver. Then he looks at silver production, silver salvage and India's silver. Back when silver was at $1.65 per ounce, Mr. Stahl projected it would top $2.50 and soar to new records—a projection amply fulfilled.

PHILIP M. LINDSTROM, investment manager of Hecla Mining Company, one of America's biggest silver producers. With 31 years' experience in the mining business—as geologist, mine engineer, mine operator and analyst—he discusses silver from the miner's point of view. He talks about the likelihood of developing new silver mines; the labor situation; price controls; and other factors that affect silver production. Then Mr. Lindstrom projects the trend in silver production, consumption and prices over the next few years.

WALLACE HANSON, contributing editor of *Popular Photography*, answers affirmatively to the question: Is silver essential in photography? He describes the rapid growth of the photography industry—the largest single consumer of silver—and shows why a substitute for silver is light years away.

DR. PAUL C. HENSHAW, president, director and chief executive officer of Homestake Mining Company—the biggest gold producer in the Western Hemisphere. He talks about the skyrocketing popular interest in gold; reviews commercial gold consumption; explains why gold production is in a downtrend; shows what it's like operating a huge gold mine; and in conclusion, explains his outlook for gold prices.

HARRY BROWNE, author of *How You Can Profit from the Coming Devaluation* and *You Can Profit from a Monetary Crisis,* both runaway bestsellers. He explains why he believes America's worst depression has already begun. Mr. Browne reviews the advantages and disadvantages of several gold and silver investments, foreign currencies and foreign bank accounts.

Taken together, these gentlemen provide you with a solid grounding in economics, gold and silver. I'm pleased to commend them to you now.

Long Beach, California
May 1974

1

Why Inflation Must Lead to Recession or Depression

Murray N. Rothbard

Dr. Murray N. Rothbard is an economic analyst, scholar, and prolific author. For 27 years he has studied economics with special emphasis on the workings of free market capitalism.

Dr. Rothbard received his Ph.D. in economics from Columbia University. In postgraduate studies at New York University, he studied with Professor Ludwig von Mises, economist and author of Human Action. *Dr. Rothbard's theoretical work developed from the logical, laissez-faire approach successfully advanced by Mises. Today Dr. Rothbard is professor of economics at the Polytechnic Institute of Brooklyn.*

Dr. Rothbard is author of America's Great Depression; Power & Market; Man, Economy and State; The Panic of 1819; *and* What Has Government Done to Our Money? *His new book,* For a New Liberty, *has been published by Macmillan.*

Dr. Rothbard has contributed articles to In Search of a Monetary Constitution, On Freedom and Free Enterprise *and four other books. His articles and interviews have been published in the* New York Times, Intellectual Digest, The Freeman, Human Events, Quarterly Journal of Economics, American Economic Review *and dozens of other publications, both popular and scholarly. He is the editor of the* Libertarian Forum, *a monthly newsletter published in New York City.*

The Pacific Coast Coin Exchange's (PCCE) exclusive interview with Dr. Murray N. Rothbard took place in his New York City residence and was published on November 30, 1972. Our interview opened with a discussion of two problems that have long plagued investors—inflation and depression.

PCCE: Professor Rothbard, in your book, *America's Great Depression,* you say that the business cycle of booms and busts doesn't happen randomly; instead, one phase of the business cycle follows almost logically from the other phase. What's the cause of this?

ROTHBARD: That's an important point to stress, because most economists today think of the parts of the business cycle as being unconnected with each other. We have a boom. We have a depression. These come from on high, so to say, then the government does something to correct them. There's no understanding, among most economists, that the business cycle is an interconnected whole.

Essentially, the business cycle is caused by inflation. Once you have an inflationary boom, then recession or depression becomes inevitable. There's no way of getting away from it.

INFLATION

PCCE: What produces inflation?

ROTHBARD: An inflationary boom is triggered and fueled by bank credit expansion which, in turn, is generated by the central banking system under the control of the national government.

PCCE: How does that work?

ROTHBARD: The Federal Reserve System buys government bonds on the open market. Let's say it buys a $1,000 bond from a dealer. It gives the dealer a check for $1,000 drawn on the Federal Reserve System.

The dealer deposits the check in his bank. As soon as the deposit is made, the bank's reserves go up by $1,000. Then the bank can make more loans—the exact amount depending on the legal reserve requirement.

Right now the reserve requirement is about 17 percent. This means the bank has to keep 17 percent of the $1,000 deposit—$170—on hand. It lends out the other 83 percent or $830.

Somebody deposits that $830 in their own bank. In turn, their bank loans out 83 percent of the $830, or $688. After a while, you wind up with a $5,000 expansion of the money supply on the basis of the original $1,000.

You can see the process very simply if you assume there's one monopoly bank in the country. Let's say every bank was a branch of the Bank of the United States. In this case, the bank would put that $1,000 into its reserve and lend out $5,000, thereby keeping within the 17 percent legal reserve requirement.

So the way the government expands bank credit is to have the Federal Reserve buy bonds. It's an arcane but very important setup—a quiet process that gets almost no publicity at all because it's not a dramatic thing.

PCCE: And that's the actual forerunner of inflation.

ROTHBARD: Yes. All schools of economics agree that increasing the money supply and bank credit causes price inflation and creates an inflationary boom.

CONSEQUENCES OF INFLATION

PCCE: What happens during an inflationary boom?

ROTHBARD: You get important, but not easily visible, distortions in the production system. An inflationary boom leads to an overexpansion of long-range capital investment (industrial raw materials, machine tools, plants and so forth), and an under-investment in consumer goods.

PCCE: How does the expansion of bank credit misdirect investment into the capital goods industries rather than consumer goods?

ROTHBARD: When the government, through its central bank, stimulates bank credit expansion by increasing the cash reserves of all the national commercial banks, the interest rate is pushed down below its free market point. In other words, as the supply of bank loans increases, the interest rates decline.

The lower interest rates make long term projects which previously looked unprofitable to businessmen, now seem profitable, and they borrow more money to invest in capital goods. As a result, wage rates and prices of raw materials are bid up.

PCCE: Why is this important?

ROTHBARD: The problem comes when the workers and landlords—who received wages and rent from the businessmen—begin to spend the new money. They consume and save in the same proportions as before the credit expansion—they don't save and invest enough to finance the purchase of the newly produced capital goods. So when the government stops its bank credit expansion, as it inevitably has to do, the boom stops.

Then the unsound over-investments of the boom are liquidated, and depression follows. So the depression or recession eliminates the wasteful investments and shifts resources back to consumer goods.

PCCE: Is this why entrepreneurs who have been successful, profitable businessmen for five, ten or maybe even 25 years, suddenly go bankrupt when a depression hits?

ROTHBARD: Yes. The shift from capital goods to consumer goods leaves many businesses stuck with more capital goods than they can afford to pay for, so they go bankrupt.

PCCE: How long does this process take?

ROTHBARD: The whole process could begin and end within a period of three to four months. You might say then, well how come the inflationary boom usually lasts for many years?

The reason why the recession doesn't catch up with the boom immediately and end it naturally, is because the banks keep on inflating. There isn't just the one sharp credit expansion or one sharp inflation—it's a continuing process where the mechanical rabbit always keeps one step ahead of the dog that's chasing it.

And it's difficult to keep always one step ahead of the cost squeeze. They have to inflate more and more as time goes on, never allowing the rise in costs in the capital goods industries to catch up with the inflationary rise in prices. They increase money and credit even more than before just to keep up with the new increase in prices.

THE GOLD STANDARD

PCCE: Why can't a government just continue inflating the money supply forever?

ROTHBARD: The reason depends on whether or not you have a gold standard. In the old days, while we were still on the gold standard, the banks could not continue inflating forever, because as they'd pyramid more and more money and bank credit on top of gold, the public would get scared. And as the public got scared, it would start withdrawing gold from the banks.

Moreover, we'd have a balance of payments problem. The inflated credit would lead to higher prices at home and to greater incomes. People would spend more on imports. Exports would fall off because of our higher prices and gold would start flowing out.

So you would have an increasing pyramid of credit on top of a decreasing gold base. The banks would eventually get into deep trouble. They'd have to stop and then contract their credit.

PCCE: Is that what happens in a bank panic?

ROTHBARD: Yes! Exactly—deficits in the balance of payments. The banks and government panic, because how do you retain gold redeemability until you stop increasing credit. Then there are runs on the banks. The result of all this is a contraction of the money supply and a decline in prices.

Incidentally, this is what happened in the late '20s. The 1929 depression was triggered by a consistent deficit balance of payments—a constant outflow of gold from the American banking system.

Now that we are off the gold standard, this constraint no longer exists. The balance of payments constraint hasn't existed since August 15, 1971, when President Nixon repudiated the gold obligation altogether.

But coming back now to your previous question, when you don't have a gold standard there's still one final restraint on unlimited eternal bank credit expansion—that's runaway inflation. If you keep up the acceleration of the money supply this will lead to an accelerated and runaway inflation of prices. This has happened to many countries in the twentieth century.

Runaway inflation, of course, means a currency collapse and chaos.

RUNAWAY INFLATION

PCCE: Is runaway inflation a possibility in America?

ROTHBARD: Yes, and as a matter of fact, that's one of the reasons why the Nixon Administration put the lid on the money supply for a while. In the 1950s, prices were going up by something like one to two percent per year, which doesn't sound like very much. During the Kennedy Administration prices were going up about three to four percent per year and during the Johnson Administration prices went up about six percent per year—sort of an exponential increase.

PCCE: Also, Professor Rothbard, in the 1960s the money supply was growing at about 3.4 percent per year. In 1970 it went up to 5.4 percent. In 1971, it increased to six percent and this year it looks like its going to increase between eight and ten percent.

ROTHBARD: Yes, it's around eight to ten percent now.

PCCE: Then what you predicted about the money supply increasing is happening. So runaway inflation, which happened previously in America—

ROTHBARD: Yes, runaway inflation has already happened several times in America. The first time was during the Revolutionary War when the American Revolutionary Government put out the so-called Continentals—paper money which depreciated astronomically. Towards the end, something like a thousand Continental dollar bills were worth only one gold dollar. Finally, the Continentals disappeared altogether, which is the origin of what used to be a famous phrase: "Not worth a Continental." The result of that runaway inflation was a healthy distrust of paper money which lasted for a long time.

The next big episode was the Civil War. In the South they had a runaway inflation.

In the North, there wasn't a runaway inflation, but the greenback went down to something like a third. In California, for example, where there was a lot of gold, it was used as money throughout the Civil War. There you had a situation where a greenback—a paper dollar—was down three to one. Prices and goods were triple in terms of greenbacks, and still the same in terms of gold. This makes it obvious that the cause of inflation is not unions or businessmen or greedy speculators or whatever, but the government which has issued too much money.

WHY UNHAMPERED CAPITALISM IS DEPRESSION-FREE

PCCE: Then the boom and bust cycle is not really part of the free market.

ROTHBARD: No, certainly not. In a free market, you have fluctuations of various sorts, for instance seasonal. An example I like to use is the seven-year locust cycle. Locusts come, say every seven years. So every seven years there's a boom in locust-fighting equipment. People use it, but after the seventh year is over, the locust-fighting equipment business is phased out.

And you may have other types of fluctuations—a decline in horse-and-buggy business replaced by a big increase in automobile business, and so forth. But you don't have an overall business cycle.

PCCE: But, Professor Rothbard, why can't individual banks simply expand credit?

ROTHBARD: Suppose one bank—the Rothbard Wildcat Bank of Northern Minnesota which, say, I own—decides to issue paper money and new bank deposits or whatever. The Rothbard Wildcat Bank will lend them out because this is how the bank makes its profits.

Shortly after I lend them out, my customers will take this new money—either paper money or bank deposits and buy equipment, food or something. They'll buy from clients of some other bank. As soon as they do that, the other bank will call upon me for redemption.

I'll have to pay up in cash—in gold, silver or whatever the monetary standard is. If I don't have the money to redeem it, I'll go bankrupt.

That salutory check of bankruptcy keeps individual banks on the free market from expanding credit very much. However, if there's a Federal Reserve System or Bank of England around to supply individual banks with new reserves and to coordinate the inflation, this is when you will have an inflationary credit expansion.

WHY THE GOVERNMENT FOLLOWS INFLATIONARY POLICIES

PCCE: Why does the Federal Reserve System expand bank credit?

ROTHBARD: The government needs money to pay its bills. It's running a deficit, which of course increases the national debt. It doesn't want to raise taxes, so it engages in deficit spending.

Whether or not the deficits are inflationary and lead to business cycle distortions depends on how they're financed. There are three ways.

First, the government could simply print money and spend it. This would be inflationary—it would raise prices. However, (a) it would not cause the boom and bust business cycle because it would not be loaning money out to business and thereby distorting interest rates; and (b) there would be no increase in taxes.

Second, is for the government to sell bonds to the public. If you or I bought the bonds, we'd simply write out a check on our bank. The Treasury would get the check and spend it on missiles, paper clips or whatever. So the money would circulate but would not increase. This would not be inflationary. However, it would mean the taxpayer would have to kick in and pay the money back plus interest. So it would increase taxes. The government doesn't like this alternative because they have to pay a higher rate of interest to us than to the banks who are, in a sense, creating money out of thin air and therefore charge a lot less.

So instead, our government chooses the third method—selling bonds to the banking system—which combines the worst features of both. It raises taxes because the public has to pay interest to the banks. It's also inflationary and causes the boom and bust business cycle.

But financing the national debt is not the only reason why the government expands bank credit. It also does so in an attempt to smooth out the ups and downs in the price level. The idea here is to manipulate the money supply by taking out money when prices are going up (and thereby lowering prices), and pumping money in when prices are falling (and thereby propping up prices).

First, keeping the price level constant is an erroneous objective. It's an inflationary objective because an unhampered capitalist system would result in a steadily falling price level—and it always has—once inflationary bank credit is eliminated from the picture.

Then, in practice, the Federal Reserve System simply can't

manipulate the price level accurately. There's a problem of statistical lag; a time and information lag. For example, suppose there's a recession in November of 1972. There are bankruptcies and prices are falling. First, it takes about two months to gather the statistics and ship them to Washington. So it's about January or February before the government finds out that the prices were falling in November. Then the government has to decide whether this is just a temporary dip, or a trend. That takes a few more months to figure out, and brings us to April or May. Then the government needs another few months to decide what to do about it.

By this time, you're into July or August. Finally the government starts doing something. Let's say it pumps money in to offset the recession. It takes another few months for the impact of the August decision to be felt in the economy—and that's around next November.

Already a year has passed—you might well be in another phase of the business cycle. Quite likely the prices are going up, so all the government's action will do is aggravate the inflation.

Nevertheless, the Federal Reserve System has assured bankers that we could always stabilize the price level. Indeed, the price level was stabilized in the 1920s. As a result, many economists said there's no inflation and therefore there can be no depression.

One of the very few economists who predicted a depression during the late '20s was Ludwig von Mises. What Mises said was that this seemingly noninflationary situation was deceptive; that there would have been a fall in prices if not for the inflationary credit expansion in the United States and Europe.

Because of this inflationary credit expansion, he predicted there would have to be a bust really soon—and that there would be a severe recession or depression. As we all know, Mises was proved right.

PCCE: Are there parallels between then and now?

ROTHBARD: Yes, definitely—there are *ominous* parallels.

PARALLELS BETWEEN 1929 AND TODAY

PCCE: In addition to overoptimism about the country's economic health, what other parallels do you see between now and 1929?

ROTHBARD: For a clear understanding, we should first examine what caused the Great Depression of the 1930s, and then examine what's happening today.

Let's start with World War I. During the war, the countries of Western Europe inflated their currencies to finance the war effort. To do this they went off the gold standard. This lead to competing devaluations, exchange controls and protective tariffs. The monetary system was chaotic and the whole international trade picture was shot.

So after the war the problem was how to recontruct the monetary system.

The intelligent thing would have been to go back to the gold standard at the then current pars and start from there. But the British insisted on going back to their prewar par which was something like 60 percent higher.

Going back to the pound at an overvalued par created a deflationary situation. It meant that British prices were no longer competitive in world markets—yet Britain lives by foreign trade.

The classical solution would have been deflation—contracting the money supply and lowering prices. The British couldn't do that because they were heavily unionized—and the unions were against deflations.

The only thing left for the British to do, to avoid an outflow of gold, was to induce other countries to inflate along with them. In the end they induced the United States to inflate the dollar and keep American prices up.

So at various times from 1921 to the crash of 1929, the Federal Reserve inflated the money supply by buying bonds heavily in the open market. This pushed interest rates below the free market level, and inflated the money supply by about 50 percent. (The rate of inflation is substantially faster today than in 1929.—Editor)

The inflationary boom led to the inevitable overexpansion of long-range capital investment, and underinvestment in consumer goods. Stock market and real estate prices boomed. Yet, this inflation was masked because wholesale prices were stable. The result was the stock market crash of 1929 and the beginning of the worst depression in modern history.

So, to come back to your previous question, we see three important parallels between 1929 and today: government expansion of bank credit and the money supply; stock market and real estate booms; and balance of payments deficits.

HOW WELL-INTENDED GOVERNMENT POLICIES PROLONGED THE DEPRESSION

PCCE: By 1932, hundreds of thousands of businesses had gone bankrupt. One out of every four workers was unemployed. Why was this depression of the 1930s so severe and so much more extended than previous depressions?

ROTHBARD: It was bound to be severe because of the large inflation that had taken place. But in 1921, we started off with a depression just as severe, and it was very short. Why, then, wasn't the Great Depression of the '30s as short?

Well, before 1929, the policy of the federal government during an American depression was generally laissez-faire. For example, during the depression of 1921, President Harding didn't intervene in the economy. So prices fell sharply, shaky businesses collapsed, and there was heavy unemployment. But in nine months the whole thing was over.

President Hoover abandoned this sound economic policy with the Great Depression of the 1930s. When the stock market crashed, Hoover ordered the government to step in with massive public works spending, prop up wage rates and prices, increase taxes, and lend money to unsound businesses.

He called in many top industrialists in the country and

bludgeoned them with so-called "voluntarism." In effect, he said, "Keep wage rates up, and continue construction voluntarily, or I'll get Congress to force you to do it." The industrialists agreed to do it, and for the first time in the history of the world, real wage rates went up in a depression.

The result, of course, was prolonged and very severe unemployment. If the government allows wage rates to fall, so will unemployment. But if the government insists on higher wage rates, then there will be more unemployment. And that's exactly what resulted from Hoover's policy of intervention.

As these government policies took us deeper into the depression, the government only intervened further. By the time Hoover left office in 1933, production had fallen 50 percent. Unemployment reached an unprecedented 25 percent of the labor force.

The capital goods industries, which had been the first to profit from the inflationary boom before 1929, were the hardest hit by the depression. Business construction fell 84 percent. Factory employment dropped 42 percent. Pig iron production decreased 85 percent. The value of construction contracts fell 90 percent. Industrial stock prices declined 76 percent.

PCCE: What effect did Roosevelt have?

ROTHBARD: When Roosevelt took office in 1933, he tremendously expanded Hoover's interventionist "New Deal" policies. The result was the longest and most severe depression in American history.

In all previous depressions, the price level fell. The banks had to contract credit to stay afloat. This brought about a decline in the money supply which led to a general drop in prices. For example, from 1929 to 1933, there was a general drop in prices. Yet, from 1933 to 1937, even though we had a deep depression, prices were going up.

INFLATIONARY RECESSIONS

PCCE: Why were prices going up?

ROTHBARD: In every depression until the Great Depression of the 1930s, the shift of resources from capital goods to consumer goods caused capital goods prices to fall much more rapidly than retail prices. During 1933 to 1937, however, Roosevelt was inflating the money supply, so the price level could not drop.

Since the price level couldn't drop, the shift from capital goods to consumer goods caused consumer goods prices to go up faster than capital goods prices. This means the cost of living went up—we had a simultaneous depression and inflation.

This happened again in 1969 when the Nixon Administration attempted to cut back accelerating inflation by tightening credit. The result was a necessary recession in the capital goods industries. As the recession set in, the administration panicked and around February 1970 they turned the money tap on again.

But once a recession takes hold, it's not easy to get out of it. When the Federal Reserve pushed interest rates down again, it increased inflation, but the capital goods industries were still in a recession. This is why we had an inflationary recession up until this year.

Nowadays, a fall in the money supply is considered unthinkable by economists and politicians. As a result we may never again have an actual fall in the price level. Instead we now have both recession and increases in the cost of living at the same time. The consumer gets it in the neck both ways.

The government should not try to prop prices up; not try to keep wage rates up; not try to shore up unsound businesses. It should allow the liquidation process—the necessary market adjustment—to take place. Then the recession or depression is over with quickly, instead of lingering on and on as in the Great Depression.

When we finally got out of the Great Depression, it was despite the New Deal rather than because of it. We got out of it by World War II. We had ten million unemployed, and then ten million men went into the army. That eliminated unemployment.

HOW WAGE AND PRICE CONTROLS LEAD TO SHORTAGES, RATIONING AND BANKRUPTCIES

PCCE: Back in the '30s, the government tried to end the depression by keeping wages and prices *up*—with the harmful results you just mentioned. Today, the government is trying to stop inflation by keeping wages and prices *down*—

ROTHBARD: Price controls cannot stop inflation. Price controls can only make things worse.

The way they work is very simple. Let's say the price of Wheaties is 15¢ a box, and for some crazy reason the government orders that there can be no Wheaties sales above 5¢ a box.

You can bet very few people would keep producing Wheaties. But lots of people would clamor to buy them because Wheaties would be a big bargain at a nickel a box. The end result would be an enormous shortage in Wheaties—an excess of demand over supply. Nobody could find Wheaties anywhere. Then you'd have a black market in Wheaties with "hot" Wheaties sold on the street corner to passersby.

Price controls fool a lot of people because the day price controls are imposed, nothing seems to change. This is because the price control inevitably freezes prices at the existing level, which is somewhere around the free market price. The next day's equilibrium is about the same too, so nothing much is going to happen at first.

It takes some months for the distortions to pile up. Then as time goes on, there will be more and more distortions, Shortages, black markets, and rationing will develop gradually—then accelerate.

PCCE: Why do many businessmen support wage and price controls?

ROTHBARD: One reason is many businessmen believe wage controls will be more effective than price controls. Therefore wage rates will be kept down. They think this is to their advantage.

But just as price controls below the free market price bring about shortages, so wage controls below the free market rates bring about labor shortages. And labor shortages have some very unfortunate effects.

For example, suppose Firm A is paying a worker $10,000 a year, and a competing firm, Firm B, is paying $12,000. The worker at Firm A naturally has an incentive to shift to Firm B. The only way that Firm A can keep him from leaving is to raise his salary to $12,000.

But all of a sudden the government freezes wages. They tell Firm A, "You can't raise the guy's wages."

The result is that many firms lose their best employees. This leads to increasing bankruptcies, especially among the firms that were just about to raise wages, but were prevented from doing it by the freeze.

So either you have increasing bankruptcies, or many workers will be forced to remain on their jobs for the duration of the inflation emergency. This may sound far out, since it's a form of slavery, but it's very realistic.

PCCE: Has this happened before?

ROTHBARD: Yes. For three years after World War II, the American Army of Occupation imposed a severe system of wage and price controls on the German economy. It was done deliberately to smash the German industrial economy—to suppress, cripple, and punish them.

The result was severe shortages of necessary products, along with a severe labor shortage. Labor was then rationed in Germany so that no German worker could get a job except through the American Army Labor Exchange. The American Army told German workers where to work. They said, "All right, we'll assign you to a Hamburg steel plant." If the fellow didn't want to work in a Hamburg steel plant, then he couldn't work at all.

This is the logical conclusion of rigidly enforced wage controls, and we'll see it here if this keeps up.

PROFIT CONTROLS

PCCE: How about profit controls?

ROTHBARD: They're horrendous. With profit controls, if

you make high profits you are forced to lower your prices, and if you make low profits you can raise your prices or keep them the same.

Any imbecile should be able to recognize their consequences. Business firms will lose their profit incentive. Businesses making high profits realize the price commission will force them to lower their prices, so they'll just find ways of not making profits. Profit controls give businesses a direct incentive to be inefficient, instead of being efficient, as they would be in a free market.

HOW THE NATIONAL DEBT HARMS THE ECONOMY AND DAMPENS STOCK PRICES

PCCE: Turning to another current problem, Professor Rothbard: the national debt is now about $450 billion. That's more than ten times what it was before the Great Depression. This year's budget deficit is going to hit $35 billion, and many economists are now saying that the budget can't be balanced in the foreseeable future. What effect is this going to have on the economy?

ROTHBARD: In the first place, the interest on the national debt will go higher. During the 1930s and '40s, we heard from Keynesian economists that it doesn't matter what the national debt is, "because we all owe it to ourselves."

But we don't owe it to ourselves. Some of us owe it to others of us, and it has to be paid for through taxes. The tax burden, of course, is a big drag on the productive system, so the national debt becomes an evergrowing burden.

Second, deficit spending usually is financed through bank credit expansions. So it inflates the money supply, which is now going up eight to ten percent a year. This is a very, very high rate and will cause greater inflation next year than we have now.

A third unfortunate consequence of the national debt is its effect on the stock market. The Treasury sells bonds in the bond market to finance its increasing deficits. This pushes up interest rates. As inflation proceeds, and as the Treasury has to finance more bonds, interest rates keep going up.

This increase in interest rates means there's pressure on the bond market, and a damper on the stock market.

PCCE: Would you clarify this, please? On one hand, we have inflation which, according to theory, should drive bond prices up and interest rates down. Yet the interest rates are going up.

ROTHBARD: Well, what happens is, initially, credit expansion pushes down interest rates below the free market level. However, as inflation picks up momentum, as prices start going up, interest rates go up to offset this increase in prices.

For instance, if the interest rate is six percent, and prices go up six percent, we essentially have a zero interest rate. This wipes out the creditor's gain because the money he gets back is now worth six percent less.

As a result, creditors will raise interest rates to compensate for the price increase. So as the inflation takes hold and really gets going, and people realize it's going to continue—interest rates inevitably go up.

It's one of the big reasons why increased profits and inflation don't necessarily raise stock prices. The interest rate offsets it.

BALANCE OF PAYMENTS DEFICITS

PCCE: In addition to the domestic problems you've been describing, America has had chronic balance of payments deficits for the last 10 years—and it's getting worse. What causes them, and what is their consequence?

ROTHBARD: The basic reason for our balance of payments deficits is the American monetary inflation. This raises prices and makes American goods less competitive in the world market.

Basically, the dollar is now the foundation of all credit—and we've been inflating the dollar for more than 30 years.

As a result, gold started flowing out because of the balance of payments deficit.

As gold flowed out, the American dollar became shakier and shakier. We had to keep persuading, inducing, and coercing other countries to pile up dollars and not redeem them in gold.

INTERNATIONAL TRADE WARS AND DEPRESSION

PCCE: How did we prevent foreign countries from cashing in their dollars?

ROTHBARD: By all sorts of methods. Our expansionist foreign policy, for example, gave us leverage. We threatened to withhold aid if the other countries redeemed their dollars for gold. But as they kept piling up more and more dollars, and as our gold supply grew smaller and smaller, these foreign countries got edgier and edgier.

Eventually, on that black day of August 15, 1971, the Nixon Administration was forced to repudiate the dollar and declare a national bankruptcy. When the United States did this it was saying, in effect, "Even though we had declared our obligation to redeem our dollars in gold, we are no longer going to do it, period." "We closed the gold window" is the polite way of saying it.

This has now plunged us into a system like the 1930s. But this is worse, because at least in the 30s the dollar was still redeemable in gold. Now, with a completely fiat system, anything could happen. We'll have a harsher system of currency blocs, competitive devaluations, and economic warfare.

We might have a dollar bloc, a sterling bloc, a gold bloc in Western Europe—and maybe even an Asian bloc of some sort.

West Germany, Switzerland, Britain, France, etc., rely very much on international trade. If we have a breakdown of international trade and international investment, due to currency blocs, devaluations, tariff blocs, etc., these countires would suffer a severe depression—no question about that. Severe depression in these countries will result in depression at least in American export industries.

WORLDWIDE RUNAWAY INFLATION

PCCE: Western Europe seems to be taking very definite steps toward an international monetary union based on gold. Do you think there is any chance they can accomplish their aim?

ROTHBARD: Almost none. The entire United States establishment, both Democratic and Republican, stands almost hysterically opposed to gold.

The United States, being an inflationary country, wants to get rid of gold altogether. The reason the government hasn't been able to do this so far is because the hard-money, Western European countries are totally against it.

What the United States would like is a World Central Bank, controlled by the United States, with a new paper currency. Then, if we have a balance of payments deficit, instead of having to pay up in gold we can pay up in Special Drawing Rights (SDRs).

So far, SDRs have been very limited. We have only a few billion of them. But with the American objective of a World Central Bank, controlled by us, issuing SDRs at will, we can simply have SDRs issued to pay our debts. We could then go merrily on our way, inflating forever, so long as we have control over the World Central Bank.

This means we'd be coordinating inflation on a worldwide scale. We could leap over the balance of payments constraint, but the result would be an eventual worldwide runaway inflation that would create chaos.

PCCE: If we should ever have runaway inflation, can commerce continue?

ROTHBARD: Well, trade continues but starts cracking up and collapsing. Runaway inflation is much worse than a depression. It's sort of like a super-depression combined with inflation.

WHAT LIES AHEAD

PCCE: So we in America are facing accelerating inflation. We have wage and price controls. We face an enormous, rapidly-growing national debt. We face the probability of trade wars. Worldwide runaway inflation is a possibility. Where does all this end?

ROTHBARD: It ends rather badly, I would say. Increasing inflation next year is almost certain. But the exact moment trade wars, runaway inflation, or depression may strike is hard to know.

PCCE: What type of occurrence could trigger this off?

ROTHBARD: One big collapse. One big West German firm, or French bank, or whatever. One healthy bankruptcy could trigger the whole thing off.

PCCE: Could it trigger off a worldwide depression?

ROTHBARD: Yes. First a West European one—and then a worldwide one.

HOW TO PROTECT YOUR ASSETS

PCCE: How can we protect our assets against depression and runaway inflation?

ROTHBARD: Well, of course it's very difficult. In general, stocks are pretty bad investments because we have high interest rates that keep a damper on stock prices. And the value of government savings bonds has been half wiped out by inflation already. The poor people who bought savings bonds in World War II bought them at low interest rates. As prices went up, their capital got wiped out.

The classical method—and I would say the best method—to protect yourself against runaway inflation or depression is with durable, highly-valued commodities like jewelry, paintings, gold, and silver.

I see a great future for gold and silver coins as the currency people may increasingly turn to when paper currencies begin to disintegrate.

PCCE: If you wanted to hedge against a possible runaway inflation, or a possible depression, would now be a good time to buy gold and silver coins?

ROTHBARD: Absolutely.

2

The Historical Perspective

Dana L. Thomas

Dana L. Thomas is a distinguished financial writer and editor with 18 years of experience on Wall Street.

A Harvard graduate in history, Mr. Thomas has written articles published in The American Scholar, Reader's Digest *and numerous other publications. In 1956, he joined* Barron's *as associate editor. In the years since then, Mr. Thomas has written more than 180 articles for* Barron's *on almost every conceivable topic, including tax shelters, computers, the monetary situation and gold.*

He has written many books, including The Plungers & The Peacocks, *a best-selling history of Wall Street, and* The Money Crowd, *which reports on many people and things making news today—including Harry Schultz, C.V. Myers, the Rothschilds, Howard Hughes, Swiss banks, inflation, silver and gold. Mr. Thomas' books have been translated into French, German, Spanish, Portuguese, Hebrew, Arabic, Hindustani and Japanese.*

This interview took place in Mr. Thomas' home in New York City and was published in October 1973. Before beginning the interview, Mr. Thomas asked us to emphasize that he is not an investment counselor and does not give advice on specific investments. We started by discussing a problem that is rapidly worsening in America—inflation.

PCCE: Would you begin by reviewing, historically, how people fared with inflation?

THOMAS: Well, throughout history, governments, to service overwhelming debts, have printed paper money—thereby wiping out the savings of their people and plunging them into misery. Time and again, we've seen that a serious erosion of money is a major herald of social collapse.

Inflation got out of hand in ancient Rome. The emperors progressively debased the currency, reducing the precious metal content in coins. So it took more and more money to buy the same amount of goods. This was a major factor leading to the collapse of the Roman Empire.

During the French Revolution, the government issued a paper money called *assignats*. For six years, from 1790 to 1796, these *assignats* lost value, and prices rose dangerously. For instance, a bushel of flour that cost a Frenchman the equivalent of 40 cents in 1790 cost $45 by 1796. A cartload of wood that cost a farmer $4 ended up costing $250. Eggs soared from 24 cents to $5. The price of a pound of soap, from 18 cents to $8.

Worst hit were people who could least afford it. The price of a comparative luxury like soap increased 4,000 percent, while an essential like flour increased 11,000 percent.

And a very natural thing occurred. On February 18, 1796, a mob broke into government offices, seized a cartload of *assignats* together with the plates and machinery that had printed them, and hauled all this into the Place Vendome. Then, at nine in the morning, before the clenched fists and curses of 300,000 people, the machinery was smashed into pieces and tossed onto a blazing bonfire of *assignats*.

To put it in a nutshell, this inflation and the chaos it produced helped bring on the dictator Napoleon.

INFLATION IN GERMANY

PCCE: Another famous example of inflation is Germany in 1923.

THOMAS: Yes, inflation plunged Germany back into the Stone Age. Eighty-four printing plants around Germany were operating day and night pouring out paper money backed by nothing more than the government's promises. At the height of this inflation, prices were rising 10 percent an hour. The mark fell to something like one *trillionth* of its value.

Inflation in Germany reached the point where it took a wheelbarrow full of paper marks to buy a cigarette. In fact, it was cheaper to light a cigarette with a paper mark than to buy a match.

Isaac Marcosson, a journalist for the *Saturday Evening Post*, visited Germany and reported on conditions there. One morning he left Cologne for Paris and bought a paperback to read on the train. He paid 72,000 marks for it. When he returned five hours later to buy a second copy for a friend, the price had risen to 180,000 marks. At lunchtime, Marcosson ordered a steak. The price on the menu was 80,000 marks. While he was eating it, the waiter informed him the price rose to 100,000 marks. By the time he was ready to pay his bill, it zoomed to 125,000 marks.

Inflation encouraged wild speculation. One of the most successful speculators—for a while—was Hugo Stinnes, an industrialist who amassed $500 million during three years of this inflation. Realizing that the mark was becoming worthless, while everyone else thought it had value, Stinnes went heavily into debt to buy companies. His obligations liquidated themselves automatically as the mark continued to fall.

At his zenith in October 1923, he owned 20 coal companies; 29 smelting works; four oil fields and refineries; 11 iron ore mines; eight chemical, sugar and paper plants; four shoe factories; nine shipping companies; and a complex of banks, holding companies, locomotive works, farms and timberland holdings. But when the inflation finally came to an end, Stinnes was $42 million in debt. He went bankrupt.

The German inflation wiped out all savings. For example, *anyone who had invested 100,000 marks in government* *bonds—$25,000 in American money—during World War I would have found himself by the summer of 1923 with exactly $2.50.*

By utterly impoverishing the people, inflation helped pave the way for Hitler.

PCCE: How about the more recent examples?

THOMAS: Yes, there were quite a few. Germany and Hungary had wild inflations after World War II. Also, China had a runaway inflation after World War II that helped usher in the Communists.

Today, runaway inflation is disrupting Argentina. The only way to stop this is deflation. But then, millions of Argentinians would be thrown out of work. You would have depression and starvation. So the inflation continues to accelerate, and you have grave political instability.

In Chile, inflation has led to almost complete chaos.

PCCE: What are the warning signs of runaway inflation?

THOMAS: The historical record seems to show that once inflation passes beyond 10 percent, it's pretty much out of control.

DOLLAR DETERIORATION

PCCE: How serious do you consider inflation to be in America?

THOMAS: The dollar has lost two-thirds of its worth since 1939. And by abandoning gold-convertibility, the government took the final lid off the amount of paper that could be issued. So today, inflation is getting worse, and unless it is somehow reversed, the dollar faces the threat of dangerous deterioration.

Indeed, a Republican administration, elected on the promise to maintain a balanced budget, has engaged in the biggest deficit-spending spree of any administration in peacetime America. In their first four years in office, the Nixon people ran up a deficit close to $90 million.

America is plagued with chronic inflation, and the historical-minded cannot forget this is one of a handful of nations that has repudiated its debts in the past century—states in the American Confederacy and the Bolshevik revolutionary government in Russia are in that select group. More Americans are beginning to wake up to something Europeans have known all along: A runaway inflation can be just as ruinous as a depression.

PCCE: The dollar—and the whole monetary system—is being racked by one crisis after another.

THOMAS: Yes, the international monetary system has disintegrated into complete chaos. The Watergate affair has had a damaging psychological impact upon the thinking of foreign dollar holders. But the main factor underlying this is inflation—which continues to rise at a record rate.

The holder of a dollar may be under the impression he has a legal claim on a tangible asset. But he hasn't, because the Fed since World War II has expanded the money supply so much, and the outstanding claims far outweigh the tangible physical assets. Today's dollar is in effect a government IOU which Uncle Sam seems determined to service by going deeper into debt. This process of servicing debt by issuing more debt could, if it continues on a massive scale (and let's hope it doesn't), break the economy.

PCCE: What effect will this have on American business?

THOMAS: The only thing that keeps businessmen in the role of supplying jobs is getting an attractive return on their money. My own feeling now is that we're moving toward a situation where businessmen can't plan too far ahead. If the crises we're going through cause enough companies to lose confidence and stop making longterm capital investments, then we're in trouble.

PCCE: How about the rest of the world?

THOMAS: I don't know of a single major currency in the world that *isn't* losing value because of inflation—do you? I think there's the real danger of a rampant, global inflation.

And I want to emphasize that we are all in the same boat together. Our major banks are heavily involved in foreign loans. Our multinational companies have investments around the world. So if there's a big collapse in one place, the impact on the rest of the world could be devastating.

STOCKS AS AN INFLATIONARY HEDGE

PCCE: What are your feelings about the stock market as an inflation hedge?

THOMAS: Few individuals have made much money in stocks during the past four or five years. In fact, since 1965 a number of industries have been very largely at a standstill—this is clear in the performance of many blue chip stocks.

Another thing, the market today is characterized by extreme price volatility. You see, the stock market historically has been an auction market with tens of thousands of individuals buying and selling. As they bought and sold, they developed a gradual consensus of price.

But now the whole play has been taken away by a relatively few powerful institutions. As a result, we have an institutional negotiating market, and the time span of price action has been tremendously telescoped. Today, if a handful of institutions rush into or out of a stock, price changes that formerly took weeks and months can happen in hours.

PCCE: Have the big institutions affected the stock market in other ways?

THOMAS: Yes, market psychology has changed as institutions have taken over from individuals. In an auction market, you had individuals using their own money. Back in the 1920s, for instance, you had big blocks of stock, too, but they were controlled by syndicate guys who had their own money on the line. If they lost, they got burned. Today, though, you have syndicates like mutual funds and pension funds where the managers use other people's money.

A fund manager can get up on the wrong side of bed one morning. "All right," he says to himself, "I'm going to dump 100,000 shares of XYZ stock. I just don't like it." This could be a pretty casual decision. His dumping of 100,000 shares of XYZ can literally swamp some poor individual out there who's holding that stock with his own money on the line. But the portfolio manager doesn't care—he may have dumped that and bought 100,000 shares of something else an hour later. This is a very tricky market for the average individual investor.

PROTECTION AGAINST INFLATION

PCCE: Then what do you see as a reliable hedge?

THOMAS: Gold and silver have offered protection against bad inflations for thousands of years. I think no economic historian would disagree with that. Gold and silver were the first universally-accepted mediums of exchange.

The Rothschilds are an interesting example. Basically they rose to prominence on their supply of gold, when the rest of the world needed it. Buffeted by wars and monetary crises, the Rothschilds possessed the yellow metal, and it proved more powerful than armies and kings. As Jews, the Rothschilds were persecuted, but their strategic control of the world's most valuable medium of exchange enabled them to rise above their adversity.

Storing up gold in Asia is a veritable style of life, as I wrote in my book, *The Money Crowd.* A tourist strolling through the twisting fetid streets of Bombay will find amidst the cow dung and beggars, gold smith shops laden with gold and silver bracelets, necklaces, religious amulets for the Hindu masses, who collect them as Americans collect insurance policies.

More generally, Europeans have suffered through constant wars, revolutions, bad inflations—situations where people's lives depended upon whether they could get out with gold, silver, jewelry and other things with intrinsic value.

3

How to Survive Our Coming Depression

Thomas J. Holt

Born in Hong Kong, Mr. Holt lived in Shanghai through the runaway inflation that devastated China after World War II. He left China in 1947—after having received a degree in economics from St. John's University, an American missionary college—and came to the United States to study textile engineering.

Although he graduated in 1950 with honors, Holt could not find any employment in the textile field. So, he took the only job he could get at the time—as a packing clerk for $30 a week. At night he continued his study of economics. "I was dead broke and could not afford going to night school," he says. "I went to the public library after work to study economics and finance, because that was my first love."

His career as an investment analyst began when he landed a trainee job with Value Line, a major advisory service. He worked his way up there for about seven years. Then, he became associated with two investment banking firms, and in 1963 helped to launch an investment advisory service for the Research Institute of America.

After having been in the investment field for close to 15

years, Mr. Holt decided to go into business for himself. He says, "I felt that most advisory services were just tip-sheets for in-and-out traders. Even the supposedly better ones were hardly ever willing to stick their necks out and tell investors exactly what to do." So in 1967, he formed T. J. Holt & Co. to offer personal portfolio management services and to publish the Holt Investment Advisory. *Since then, Mr. Holt has regularly "stuck his neck out" by making unequivocal—and sometimes unpopular—opinions of the economy and the market.*

The New York Times *says Holt "has been considered one of the great villains of Wall Street because he has fearlessly been forecasting virtual disaster for the market for several years."* Business Week *says he may qualify as the "Super-bear of the year."*

Although Mr. Holt has been decidedly bearish in recent years, he is, by no means, a "professional pessimist." While he did call the 1969-1970 and the 1972-1973 bear markets, he also correctly predicted the mid-1970 market upturn. More importantly, he has done quite well for his subscribers and clients.

Recently, one of Mr. Holt's clients wrote that his stock-broker "cannot understand how I have consistently made such large gains in my portfolio while others are now taking options on window ledges." Indeed, the Holt Investment Advisory *has made 161 specific recommendations; and despite the difficult market of the last several years, they have registered an average capital growth—excluding dividends and interest—of better than 12 percent annually. Many of Holt's managed portfolios have scored even wider gains.*

What is the "secret" of Mr. Holt's success? One thing that is unique is his "flow of funds" analysis of the stock market. As he explains: "I want to find out at any given point which groups of investors are buying stocks, and which groups are selling. Then I ascertain if the current trend with each group is continuing, accelerating or reversing. Having established that, I determine as best I can what factors motivate each group, how the factors are likely to change in the future, and accordingly, how they are likely to affect the stock market."

Another factor in his success has been his position on gold and silver—he is a long-time advocate of both precious metals as hedges against inflation and depression. And as we know, current events are confirming his judgment.

We interviewed Mr. Holt in his offices at 277 Park Avenue, New York City and published the interview on August 31, 1973. We turned first to the mounting inflation in America.

PCCE: What do you expect to be the longterm trend for inflation?

HOLT: I doubt very much that any meaningful steps will be taken by the government to stop inflation. The administration has already done all it can in the sense of limiting the fiscal deficit and imposing controls. I don't think the Federal Reserve Board will do anything constructive. In the past few years, the Fed Chairman Burns has blamed inflation on everyone and everything else, but he has not indicated any desire on the part of the FRB to slow down inflation, and I don't see why he would do so now.

PCCE: Is this why we've seen interest rates going through the roof this year?

HOLT: Yes. You see, the Federal Reserve hasn't really been tightening credit seriously. On the contrary, the Fed has allowed the money supply to increase at a record rate during the past year. The more money is created, the more bank loans become outstanding and the more bank customers are in debt.

Interest rates have been going up because the banking system, as well as individuals and corporations, are now extremely illiquid. For example, the bank loan to deposit ratio is now the highest on record—not just higher than during the 1969-1970 money crunch, but indeed higher than in 1921 and in 1929. When banks do not have enough money to lend, the only way to get more money is to offer exorbitant interest rates to attract deposits.

Likewise, when businesses are illiquid, they must pay fantastic rates to get loans just to keep themselves financially solvent.

PCCE: How high do you think interest rates are likely to go?

HOLT: Short term interest rates could go up to 15 percent. We see federal funds over 10 percent in recent days already. At some point, moreover, there will be no money available to marginal borrowers. With money remaining tight—in spite of Mr. Burns—banks will become more selective with their clients. Those individuals and corporations who do not enjoy top credit rating are usually the ones who need money the most to tide them over. And they may not get it. This is when your bankruptcy rate will increase.

PCCE: Where will this lead us?

HOLT: We will first have an economic slowdown, then more inflation partly, I suppose, because it is always easy for the government to print paper money. It is a unique power that only the government has. And politicians know that when the whole economy is so illiquid, any business contraction can snowball into a major catastrophe. Naturally, they would like to postpone that indefinitely.

But sooner or later the boomerang must come back. Politicians cannot overcome free market forces. So I think we will have more inflation for a while, then collapse and depression.

RECURRING INTERNATIONAL MONETARY CRISES

PCCE: Turning now to the international scene, what do you think are the major factors behind our recurring monetary crises?

HOLT: In my opinion, the biggest factor is the excessive amount of fiat money the government is printing. None of these dollars are backed by gold.

Businessmen in most parts of the world are still accepting and using dollars now merely because of inertia—they have been using dollars for so long. But eventually, more and more people will get out of dollars into something sounder. I believe you will see the dollar lose importance in international trade.

There are a couple of fallacies about the dollar I want to touch on here. One is that the dollar is basically sound because the U.S. economy is the biggest and strongest in the world. This is like saying a guy who makes $50,000 a year must be in better financial shape than some one making $10,000. But that "rich guy" could be spending $75,000 a year and go bankrupt. The check he writes may very well bounce whereas the $10,000-man's check is perfectly solid. Like the "rich" man's rubber check, there's virtually nothing backing the dollar nowadays. It's just a piece of green paper.

Another fallacy is that because other countries are having inflations worse than ours, our comparatively lower inflation rate means that the dollar is strong. The trouble with this idea is that the inflated dollar is the primary cause of inflation abroad. In an effort to support the dollar in currency markets, Germany, Japan and other countries had to expand their own money supply rapidly, bringing on bad inflations. This is why other countries have recently become reluctant to support the dollar.

PCCE: You wrote in *The Holt Investment Advisory* that President Nixon, by pushing for wider powers over U.S. trade policy, has "sounded the bugle for an international trade war." What would be the consequences of this?

HOLT: I think we've already begun to see some consequences now. For a while, we had tariff surcharges. We've limited textile imports from foreign countries, and we have export controls. The government is erecting these and other barriers against international trade, thereby preventing people from buying things they want to buy.

Another thing, as the dollar sinks, American business becomes more competitive in world markets. This, too, is part of the trade war, and it's very upsetting to the French, Germans, and Japanese, among others. Eventually, the U.S. will have trade surpluses, but then other countries will have trade deficits. They will then want to shore up their own trade positions with devaluations and trade barriers of their own. I think even the Japanese yen and the Deutschemark will eventually be devalued. Ultimately, all governments want their currencies to go down as much as possible—so long as they are not obliged to lose gold.

This makes it increasingly risky for businessmen to carry on their international business. After all, you can see your profit margin vanish in a sudden devaluation or change in a tariff. So, world trade will slow.

PCCE: Do you think we will see an international situation like we had during the 1930s?

HOLT: Yes, I think so. We could see the same thing almost chapter by chapter. Politicians never learn from history, unfortunately. I think we may end up with so many currencies being devalued, and so many trade barriers, that we'll see a worldwide recession or depression.

ANY ALTERNATIVE TO DEPRESSION?

PCCE: It seems that whether you look at events from the domestic or international point of view, you conclude that we're headed toward a depression.

HOLT: Yes, either worsening inflation here or an international monetary crisis could trigger a depression. You can't tell for sure which it will be, but I think an international monetary crisis is more likely to do it.

PCCE: What do you think the government will do in an effort to get us out of a depression?

HOLT: I expect there will be massive government spending and massive federal deficits. We will probably have government work forces like the WPA of the 1930s.

But I don't expect these things to work. You see, whenever you have excessive credit like we have now, there must be a period of correction. The faster it is over with, the better, because then we can begin rebuilding. I think the government's policies will interfere with the correction process and lengthen the agonizing period we'll go through.

PCCE: How will price controls affect our economy?

HOLT: Since the cause of our inflation is the government's easy money policy, the new price controls won't help any more than the last three phases did. Rather, these controls will bring more imbalances in the economy.

Besides, in the coming depression I think that prices will come down so fast that controls won't make any difference anyway.

PCCE: So you don't expect an inflationary depression, despite massive government spending?

HOLT: That's right. I think that while the government will spend heavily and increase public debt, we will see an even sharper contraction of private debt, and the net effect will be a contraction of total debt.

PCCE: How will inflation and monetary crises affect the stock market?

HOLT: They're already having an effect. We've been in a primary bear market since 1968. We didn't see the peak in the Dow Jones Industrial Average or Standard & Poors until this year. But looking at the broad-based averages that are not weighted by stocks with large capitalizations, you can see very, very clearly that the market started heading down since the end of 1968. If you ask the average investor,"Did you make money in the stock market from the end of 1968 on?" he will say "No." That's perhaps the best way to view the situation.

PARALLELS WITH 1928-1929

PCCE: Earlier this year, you predicted a stock market crash was on the way. Do you still hold this view?

HOLT: Yes, I do. Actually, many stocks have already crashed. I think the stock market today is similar to 1928-1929.

First, the big buying in the twenties came in 1928, when the market went up sharply. Then individual investors started to get out of the market, and stocks started to decline. By September 1929, more than half of the listed stocks had declined 20 percent to 40 percent already.

But the Dow Jones Industrial Average lagged behind the general market trend. It was supported by the "big money." In early 1929, it went sideways for many months. During the summer of 1929, the Dow shot up for a few weeks, turned around and headed down.

Now look at what happened in the market during the past year or two. Most stocks have been declining long before the Dow Jones Industrial Average dropped off. Also notice that for many, many months last year the Dow was locked into a very narrow trading range of 900-975. Then all of a sudden, it shot up and peaked in January of this year. Since then it has turned around and dropped.

We all know that the Dow has done better than the market as a whole, because some component stocks have been supported by the institutions. During the past year, for instance, every time the Dow dropped to 900, institutional buying would drive it back up as far as 975, sometimes for a few days, sometimes for a few weeks. Then gradually it went back down again.

Another parallel is the sharp increase in margin debt. From 1926 to 1928, there was quite a large increase in margin debt, but the increase was steady. Suddenly in 1929, margin debt surged sharply. The same thing happened in 1972. All of a sudden, margin debt increased so sharply that it was two or three times the previous year's volume.

What's the significance of this? In 1927, the public was selling stocks, institutions were buying, and speculators were also buying. In 1973, the public was still selling, institutions were still buying, but speculators on margin were forced to meet margin calls by selling. And the market this year has been dropping steadily.

PCCE: You draw a sharp distinction between individual investors and institutions. Would you elaborate on that for our readers?

HOLT: During the bull market years of the forties, fifties and part of the sixties, private individuals had confidence in the market and were buying stocks regularly. These people just bought and put away "good" stocks. It didn't matter whether the economy was booming or not—people kept putting more and more money into the market. So the underlying market trend was upward.

Temporary declines such as occurred in 1954 and 1958 were created by sophisticated investors, including institutions, that are very economy-minded. When these investors had an inkling that a recession was coming, they started selling. Conversely, when they saw a recovery on the way, they started buying, and the market resumed its upward drive. All those ups and downs happened on top of an underlying upward trend.

But then individual investors began to think differently. Particularly after the sharp market drop of 1962, they began to think in terms of possible capital losses. Indeed, Federal Reserve Board figures clearly show that private individuals have been net sellers of stocks since the early 1960s. Whether economic news is good or bad, individuals keep taking more and more money out of the stock market. By the late sixties, their withdrawals were large enough to offset institutional buying, and the primary bear market began.

Sometimes individual speculators (as distinguished from the serious investor), by rushing into and out of stocks, will exaggerate advances and declines. You can gauge the amount of speculation from the volume of margin debt. Speculators, though, don't have a lasting impact on the market. They're like surface waves that do not change the underlying ocean tide.

Now, many people think institutions are very, very powerful. And, in fact, they do account for nearly two-thirds of the trading activity on the Big Board. But in terms of actual holdings, private individuals still hold more than half of the outstanding stocks, or roughly 300 billion dollars worth. Since individuals are selling stocks on net, you have hundreds of millions of dollars of stocks unloaded onto the market month after month. If they reduce their selling temporarily, it can be offset by institutions; but overall, individuals are now selling far more than any institutions can buy.

PCCE: Why have a handful of "glamor" stocks managed to hold their own in the face of heavy selling by individuals?

HOLT: We know from the flow of funds analysis that the only major group that has been buying stocks is the pension fund industry. Because of their size, institutions must invest in companies with large capitalizations. So when all this institutional money zeroes in on a few hundred stocks, they have got to go up. Even though, say, the general public is selling more than institutions are putting in, the selling is spread across the board, while the buying is concentrated into a small segment of the market.

However, this cannot last forever. We know, for example, that individual investors have been selling at least partially because they want more income. Well, glamor stocks are offering the smallest yields right now. Disney, Polaroid, Avon—these and other companies are selling at about one percent yield or less, and are not showing much growth either. Moreover, you have to discount their future growth 20 to 30 years or more before they can have even a five percent yield.

With glamor stocks offering such small yields, the general public will now shift the selling pressure to these stocks.

Remember that individuals hold about 300 billion dollars worth of stocks. If just ten percent of that stock is sold, we are talking about 30 billion dollars worth of stocks that could be dumped on the market. No matter how big the institutions are, they cannot spend that much money to prop up the glamors.

That's why I think from now on the glamor stocks will start coming down under new selling pressures. Meanwhile, some of the depressed stocks may begin to show some recovery. Already, some mutual funds have been getting into these stocks, causing them to head up. The typical mutual fund manager would figure, "I can sell the high P/E stocks to the bank-managed trusts and use the proceeds to zero in on a low-priced stock and get a much higher percentage gain." As more and more mutual fund managers do this, we will have added selling pressure on the glamor stocks.

STEADY EROSION, OR A CRASH?

PCCE: When do you think the glamor stocks will start to crumble?

HOLT: Sometime within the next several months. We have reached a point where the glamor stocks have already run out of steam. They have not come down a whole lot yet, but they have not been going up any more either.

In the 1929 crash, the stock market dropped 40 percent to 50 percent in a few months. Today, if you disregard the weighted averages, you will find that the great majority of stocks have already gone down 50 percent from their 1972-1973 highs. What we are waiting for now is the glamor stocks to join the rest of the market.

I expect depressed stocks will go up in a bear market rally, and then at some point they will go down again. Ultimately, just about all stocks will go down again. Only the precious metal issues can buck the trend.

PCCE: How low do you think the stock market is likely to go?

HOLT: I think yield is a good guideline, because this is what many people now use to compare stocks with bonds and savings accounts. I believe yields of high-quality stocks have to go up to the six percent to seven percent area, and that means the Dow Jones Industrial Average will go down to the 500 or 600 area.

PCCE: When do you expect stocks to hit that level?

HOLT: If panic comes, it can happen late this year or early next year. Otherwise, probably in the mid-1970s. You have to play by ear to see how fast the liquidation goes. The stock market could just keep on eroding for two or three years or it could crash in a matter of months. Certainly this crash will help bring on a deep recession, probably a depression.

PCCE: What is your opinion of the effects of institutional investors on the stock market?

HOLT: Very upsetting. One of the reasons the stock market is so unstable nowadays is the domination of the institutions. At the outset, though, I should say that institutions do not have to be a negative factor in the market. During the 1950s, a typical mutual fund, for instance, would take as long as two, three or more months to establish a position in a stock, buying perhaps a few hundred shares a day. If the stock's price was strong in a given day, it would suspend buying. But when the price was temporarily weak, it would buy a little more. This policy made the funds a stabilizing factor.

But in the past several years, even the old-line institutions have been acting like the go-go funds. When institutions regularly trade 2,000 or 3,000 shares of stock in half an hour, or 100,000 shares in a day, they become very disruptive.

But despite their major influence, the institutions cannot overcome free market forces. I believe we are coming to a point where they must lose control of the market even for their favored stocks. This is because no matter how much they try to support IBM or Xerox, individual investors will be selling these overpriced stocks especially heavily. Of course, the public will continue to liquidate their other holdings as well. The fact is that the average investor doesn't care much about earnings growth anymore. He doesn't like stocks now because they don't yield enough. So I think the market will continue trending downward, interrupted occasionally by bear-market rallies.

Once the economy goes bad, moreover, the growth-minded investors, particularly the institutions, will join individual investors as net sellers. That's when the real damage to stock prices will be done.

I'd like to emphasize at this point that I don't think people should give up investing. Surplus capital should always be put to work. And, I think people should always consider the market, whether it is going up or down, to be a vehicle to build capital. In a bear market, you just put a little less in stocks and you use bear-market investment strategy. Our clients, for example, have generally made good money in the market in the last few years.

THE FUTURE FOR PENSION FUNDS, BANKS, MUTUAL FUNDS AND STOCK BROKERS

PCCE: How severely will this selling pressure affect pension funds, banks, mutual funds and stock brokers?

HOLT: Pension fund money managers will be in a dilemma. They are working for trustees who now know that even "good" stocks have not gone up for a while. Just two or three months of declining, okay. But these stocks have gone sideways or downward for two or three years now. Some trustees are already starting to ask, "Look, why are we buying stocks with one percent or less yield, while our annual requirement for the trust is many times that? Why do we keep holding these stocks in our portfolio?" So the trustees will soon be telling the money managers to stop supporting the glamor stocks, stop throwing good money after bad.

Once a few large pension funds attempt to dump their stocks, the market will collapse. Then, it will be everyone for himself. Pension funds must regularly earn four percent or five percent or more a year. So if their portfolios, which are loaded with high P/E stocks, go down 25 percent or 30 percent at a time, they will be seriously set back. It will take perhaps ten years of ultra-high return to make that up, which is almost impossible. As a consequence, many pension funds, for all practical purposes, will go bankrupt. Their accrued or potential liabilities will far exceed their assets.

PCCE: And banks?

HOLT: They'll also be affected very seriously. Once some of these pension funds get into trouble, I won't be surprised if there'll be a tremendous number of lawsuits against the banks responsible for the pension fund trusts.

After the market has come down, it'll be very difficult for a money manager to justify before a judge or a jury why he —as a fiduciary—bought stocks that offered one percent yield or less and selling at exorbitant P/E's. As a result of liabilities stemming from such lawsuits and also because most banks are terribly illiquid now, I forsee multiple bank failures.

PCCE: The mutual funds redemptions have been exceeding sales. How severely is this affecting the stock market?

HOLT: This, of course, has already depressed stocks. Many mutual funds have had to sell their holdings to meet redemptions. Actually, what mutual funds did in the past several years is just a preview of what pension funds are going to be doing in the coming years.

Mutual funds started buying stocks with thin markets during the go-go days of 1967-1968, and they created tremendous performers to show shareholders. However, in going after this kind of artificial performance, the mutual funds set the stage for their own downfall; because once the public started redeeming their mutual fund shares, the funds had to sell their stocks. But we haven't seen the worst yet. So far, the funds have sold some of their more marketable holdings. Eventually, the funds have got to liquidate those stocks with hardly any market. Once this gets going, prices will really nosedive.

PCCE: What trend do you anticipate for mutual fund redemptions in the years ahead?

HOLT: Mutual fund redemptions will almost certainly continue. It slowed down for a couple of months this past summer, mainly because stock prices had come down so much—the public typically responds more to the price level than to any economic news. But looking two or three years ahead, I think mutual fund redemptions can't help but exceed sales by widening margins.

PCCE: Many stock brokerage houses are already shaky now. Will they be able to survive?

HOLT: I think stock brokerage firms are more illiquid today than they were two or three years ago during the last crunch. So we will see many more failures, mergers and acquisitions.

GOLD IN A DEPRESSION

PCCE: With the stock market heading down, many investors are looking into gold and silver. Turning first to gold, how do you evaluate the two factors affecting its price: its supply-demand situation and its value as a hedge?

HOLT: I can't help but believe that the price of gold will keep going up in the coming years, because the supply-demand situation is very lopsided.

The commercial demand for gold is increasing—I think it is common knowledge by now. And by the mid-1970s, commercial demand alone will be consuming total gold production.

Yet the hoarding demand for gold is far more important than commercial demand. Gold has stood the test of time. It never corrodes. It can be stored or hidden easily. Gold is easily portable and accepted by people everywhere as the universal store and standard of value. You can go back thousands of years and see that the price of gold has done fantastically well during inflation and depression.

I don't think that people who buy gold for hedging purposes will sell it until the whole currency situation is straightened out, and I don't expect that any time soon. Some people ask, "What about all the gold hoarded in France, Asia and the Mideast?" I'm pretty sure that the price of gold could go to $200-$300 per ounce, and most hoarders would still not sell their gold heirlooms, bullion bars and coins.

PCCE: We've heard that central banks may be planning to sell some of their gold on the open market. What do you think of this?

HOLT: I don't see central banks selling any gold as a group, though we've heard about this many times. Central banks would sell gold only if there were a concerted agreement with every major country selling at the same time—something like the defunct London Gold Pool. Stupid as central bankers are, I don't think they're stupid enough to let somebody else get hold of their gold. And even if they did sell gold, many common people around the world would be eager to buy it all up.

PCCE: What is the potential for gold consumption in Japan?

HOLT: I think it is very strong. To begin with, America

played a dirty trick on Japan. For years, we told Japan not to convert their dollars into gold. Most European countries at least had some of their reserves in gold, but Japan didn't. So practically all of Japan's reserves have been in dollars.

Then all of a sudden, we closed the gold window and started devaluing the dollar. Next, we shocked Japan with our new China policy. Now, with our many trade barriers, the Japanese are being treated like second-rate partners. So I think the best thing the Japanese could do to protect themselves is to increase their gold holdings. That's why they have let their own people buy gold. The Japanese can buy it even in department stores and supermarkets.

PCCE: What is the potential demand for gold in the future?

HOLT: Let's look at it this way. Of the entire world's population that can afford to buy gold, only a tiny fraction has bought it so far. I am convinced that with the money printing machines running at full throttle and with economic chaos and depression in prospect, a lot more people will want to preserve their savings by buying gold. Remember, these people will be buying gold not as a speculation but as insurance.

Consider this, U.S. citizens alone have total financial assets of some two trillion dollars. If Americans decide to invest only one percent of that—or 20 billion dollars—in gold, you will have many years of world gold production bought up just like that.

PCCE: Representative Henry Reuss, Chairman of the International Economic Subcommittee, said that people who buy gold are "suckers." What's your reaction to this?

HOLT: We have heard this kind of thing for the longest time. So far as these "suckers" are concerned, they should be very happy. They have protected their savings from inflation and the dollar devaluations better than anybody. I think gold can continue going up for a long time before it reaches its peak, so the "suckers" will be laughing all the way to the bank.

LEGAL OWNERSHIP OF GOLD BULLION

PCCE: Do you think that Americans will again be allowed to own gold bullion lawfully?

HOLT: Well, the Congress has already passed the necessary law, but it seems that the President will eventually be the one to decide when. So, I think it's only a matter of time before gold ownership becomes legal here.

PCCE: What effect do you think this will have on the price?

HOLT: There will be a mass rush into gold, because there simply isn't enough to satisfy the demand. My conclusion is that after we have monetary chaos and depression, and once people are allowed to buy gold lawfully, the gold price can rise faster than it has during the past two years.

PCCE: How high would you say the price of gold could go in the next few years?

HOLT: This is strictly crystal ball gazing now. Assuming most countries continue to inflate their money supplies, I would say that gold could go to $500 per ounce by the end of this decade. Even if depression causes a contraction in the money stock, I still foresee a price of $300.

PCCE: Then you would say that now is a good time to buy gold coins?

HOLT: Yes, certainly.

MASS SCRAMBLING FOR SILVER

PCCE: Turning now to silver, do you expect the uptrend in silver consumption to continue too?

HOLT: Sure. Here, too, there's a lopsided supply-and-demand picture. And there's simply no substitute for silver in sight. Photography is one of the biggest users of silver. Companies like Kodak have tried very hard to find an economical substitute for silver, and have not come close. Silver consumption in electronics and aerospace has increased rapidly.

PCCE: Do you believe U.S. silver production will increase or decrease this year as compared with last year?

HOLT: Generally, silver is produced as a byproduct of copper, tin, lead and zinc. So increasing silver prices do not necessarily mean an increase in mine production. In particular, this year, it will not be much higher because of the Sunshine Mine strike.

PCCE: In 1972, world silver consumption surpassed mine production by 179 million ounces. How do you think this will change in the future?

HOLT: I think we will have an almost unfillable gap between demand and new production. For the longest time, the gap was filled by the Treasury Department's hoard, until it almost ran out of stock. Since then, we've been using up available supplies—you can see this in the declining Comex warehouse silver stocks. In a few years, we will have a mass scrambling for silver.

Even during a recession or depression, there would still be a silver shortage. The price of base metals would fall sharply, base metal production would fall sharply, and so would silver output.

PCCE: How would you compare the performance of silver with other investments during bad times?

HOLT: Silver is very good. I believe both gold and silver are essential hedges against inflation and depression.

PCCE: How high do you think silver could go?

HOLT: I expect silver prices to keep on climbing upward. I would say in the next two years, silver could reach $3.50 or $4 per ounce. By the end of the decade, silver could go as high as $5.

PCCE: Would you say that now is a good time to buy silver coins?

HOLT: Yes, definitely.

PCCE: Overall, what portion of personal assets would you recommend putting into gold and silver coins?

HOLT: I would say 20-25 percent.

4

Runaway Inflation in China: An Eye Witness Account

Thomas J. Holt

There's a Chinese custom where people give children gifts of money wrapped in red-colored paper on Chinese New Year—it's like people here giving Christmas gifts. My family was quite well off at one time, and our friends and relatives were too. So, every New Year, I would get quite a bit of money—it would come to $30, $50 or more, which was, of course, big loot for a child. As soon as the New Year was over, my mother would put the money in a savings account for me. Including interest earned, that account amounted to over a thousand dollars by the time World War II came along. Before I left for the United States in 1947, I tried to take the money out. No luck. The bank said they couldn't give me anything, because there wasn't enough in my account to cover the lowest denomination of the currency then in use.

You see, the smallest denomination of paper money during that period was very often $50, $100 or $1,000—depending on which particular month we're talking about. At one time or another, people would write checks or carry paper money that had so many zeros that they were all millionaires. They had to carry baskets full of paper money to buy everyday items. When that happened, the government would change currency using what we may call "reverse splits" to knock off two or three zeros. In other words, they would ask the public to turn in all their old money in exchange for a new currency. For every $1,000 bill, for instance, they would get a new $1 or $10 bill. Checking accounts were automatically adjusted. Well, after all the inflation and currency changes, my savings account simply had nothing left from its original $1,000-plus total.

I knew many family friends and relatives who were quite well off before the runaway inflation got going, but by the time I left China they were living hand-to-mouth. Their savings, too, were wiped out.

I was born in Hong Kong. When I was a very young child, my family moved to Shanghai, and that's where I grew up. The Shanghai of my boyhood years was peaceful, safe and emerging well from the then worldwide depression. No one would have dreamed that runaway inflation could happen in China, home of a proud people who basically are always prepared to do a day's work for a day's pay, and don't ever believe in handouts.

But they hadn't reckoned with the devastation of World War II. Chinese industry was wiped out by the war. With it went the government's main source of tax revenue. Public servants, both civilians and the armed forces, then became underpaid. As a result, corruption proliferated. Ethics and morals gradually disappeared. To finance its ever-increasing deficit spending, the government printed massive amounts of a paper currency. The more this "Chinese National Currency," or "CNC" for short, the government printed, the faster prices increased.

No one—from the poorest laborers to the richest businessmen—kept any CNCs for any length of time, because they lost value so fast. As soon as people were paid, they would run to the store and buy the things they needed. And if there's any cash left, they would hoard oil, rice or whatever—frantically speculating in an effort to keep ahead of inflation. Naturally, some people, experienced in speculation, prospered, while others didn't do so well.

The inflation was fantastic. I remember periods when prices were doubling monthly, weekly, sometimes overnight.

People would need a huge sack of CNCs to buy a little rice. Children had to spend perhaps $100 for a bag of peanuts. I remember the prices of school textbooks would triple from one semester to the next, so that, in terms of the currency in use, we would make "huge profits" selling used books to other students. Of course, we had to put up a lot more cash to buy the books we needed. And we used to pay tuition with a duffel bag full of CNCs. We'd spend hours at the admission office counting the stuff. Things like this you just cannot forget.

Government officials tried to reassure people by saying, "Son, the rate of inflation will slow down . . . There are no fundamental reasons for prices to go up—it's greedy businessmen who are causing the inflation." Then the government would try to step in with price controls, even though it was still printing money at full speed. As always, this led to shortages and black markets. People with some money had to pay prices far more than the controlled levels to buy things they wanted in the black market. Others would line up for hours on end, even waiting overnight to buy their allotments of rice and oil in the so-called regular market. Stores would be cleaned out in a half-hour. After a while, some store owners found it more profitable to just hold onto their inventories than to do business.

JAIL, RIOTS & EXECUTIONS

Still printing money, the government would then put a few businessmen in jail, or even execute a couple, punishing them for having created inflation for personal profits. We had lots of riots. I honestly don't know whether the riots were brought on spontaneously by people bankrupted by runaway inflation, or whether they were Communist-inspired, but we had riots day in and day out. They started in the poorer sections of major cities, and then spread to the countryside.

Inflation always hits people in the lower and middle-income markets more than the wealthy. This is because a wage-earner never got raises as fast or as much as price increases. Higher income didn't come automatically when prices rose, as happens with a person in business for himself.

Some people tried to hedge against inflation by going into the Shanghai stock market. Then manipulators took advantage of the situation, purposely promoting speculation. For a while, people bought stocks like crazy. They went up thousands of percent a year. But eventually, the manipulators pulled the plug and the stock market collapsed.

My father invested most of his money in business and real estate. He operated a chain of theaters that showed Hollywood movies, and he had done quite well—until the Communists came. When Mao crossed the Yangtze River, it was clear the Reds would take over China. So my parents packed a suitcase and fled. The one suitcase and the little money hurriedly transferred to a Hong Kong bank were all they had left—the Communists seized everything else. That's why I had to borrow money to finish my last year of college here in the States.

GOLD & SILVER OWNERS PROSPER

Many relatively wealthy Chinese people hoarded gold coins, gold bars or silver coins; their savings were preserved throughout the inflation. And I remember that even our servants used whatever extra money they had to buy articles made of gold and silver. Without any knowledge of economics or finance, these people just instinctively knew that they could always convert gold and silver items into money with which to buy things later. But if they held onto the paper CNCs, they would have lost everything to inflation.

From my experiences during this inflation and from subsequent research on this subject, I concluded, first, that as long as a government is not restricted from printing money in volume, it will do so. And regardless of what assurances to the contrary they give, prices of everything will keep going up.

Second, runaway inflation will finally end in one of two ways. In a country like the United States where hundreds of billions of dollars are invested in fixed income investments, runaway inflation will eventually collapse the bond market and all pension and insurance programs—personal savings would be wiped out. You would then have deflation and depression. If deflation is allowed to run its course, the economy may then return to normal. But if the government indulges in massive deficit spending, deflation will be followed by another round of runaway inflation.

In a country like China or those South American banana republics or Vietnam—where there is no longterm capital market, to speak of—runaway inflation would lead straight to revolution. Indeed, I think that more than anything else, inflation paved the way in China for the Communists. As far as the average Chinese was concerned, he embraced communism not because of any political beliefs, but because he simply couldn't make ends meet. The Communists won because they promised an end to inflation and its chaos. By the same token, I think runaway inflation is also the key reason why the Communist government was toppled in Chile.

I have also concluded that no amount of government pep talk, economic program, or coercion will save the public from runaway inflation. You'll have to watch out for yourself and take steps to protect your assets *before* inflation or epression gets out of control. Once inflation has reached the runaway stage or once an economic collapse takes hold, you will have missed your best opportunity to protect your assets.

Finally, the most reliable hedges are gold and silver. These precious metals keep their value, or even increase in value, no matter what may happen. And gold and silver are easily portable; wherever you go, you can always take it with you. And you won't be wiped out by a wholesale collapse of the banking system.

5

Protection Against Dollar Devaluation

Franz Pick

Dr. Franz Pick is one of the world's leading experts on currency and gold.

Born in Bohemia, Dr. Pick studied law at the University of Leipzig; monetary theory at the University of Hamburg; and inflation-devaluation theories at the Sorbonne in Paris. He wrote his Ph.D. dissertation on devaluations.

Dr. Pick's introduction to the dangers of paper currency came early in his life. "When I was born," he says, "my father took out an endowment insurance that would see me through four years of university." At maturity, the insurance company paid Pick fully, but in paper currency which was so debased that "the policy paid for only two meals."

By 1933, Dr. Pick was able to put his knowledge to profitable use. "When I heard President Roosevelt explain over the radio what he planned to do, I knew the dollar was finished." So Dr. Pick went into the currency market, sold $100,000 short and made huge profits.

During early World War II, Dr. Pick was paymaster of the French Resistance. He learned to handle contraband dealings in currency and precious metals from a "trained black marketeer." In this way, Dr. Pick helped finance the Resistance. Among his other wartime exploits, Dr. Pick secured the release of an imprisoned director of the British secret service—with some well-placed gold. For his efforts, Dr. Pick was sentenced in absentia *to death by the Nazis.*

Still very much alive after World War II, Dr. Pick started publishing Pick's World Currency Report, *which each month monitors over 96 currencies, the gold and silver markets, and other factors in the monetary situation.*

From 1951 to 1955, Dr. Pick also published The Black Market Yearbook. *Since then, he has published the annual* Pick's Currency Yearbook, *a review in depth of the world monetary situation.*

In 1960, Dr. Pick wrote United States Dollar: Deflate or Devalue? *He has also written* Common Stocks Versus Gold, 1930 to 1962; Silver: How and Where to Buy and Hold It; All the Monies of the World; The Numbered Account; *and* Gold: How and Where to Buy and Hold It. *Dr. Pick's books have been published in French, German and Portuguese, as well as English.*

His articles have appeared in Barron's, Playboy, Harpers *and other publications. He contributes a regular column to the* Northern Miner. *An annual subscription to his* World Currency Report *is $300, and well worth it to the serious hedger.*

Dr. Pick also teaches currency theory in English, French, German, Spanish and Czech. His seminars are conducted in Europe and the United States. He offers private consultations at $400 per half-hour.

This interview was held in Dr. Pick's New York City office, and was published on May 31, 1973. The walls of Dr. Pick's private office are paperd with worthless currency—"Assignats" *from the runaway inflation during the French Revolution, and "continentals" from America's Revolutionary War runaway inflation.*

PCCE: On February 12, 1973, the dollar was devalued for the second time in 14 months. What is the significance of this?

PICK: It was the third fraudulent state bankruptcy in the history of the United States. The first was in 1934. The official price of gold jumped from $20.67 per ounce to $35. The second dollar devaluation, a comic book event, took place in December 1971. The official price of gold increased from $35 per ounce to $38. Then on February 12th of this year, the dollar was devalued for the third time. The official price of gold increased to $42.22 per ounce.

The two most recent devaluations alone amount to about 18 percent. The gross total public and private debt in the United States is about $5½ trillion. **So these devaluations wiped out more than a trillion dollars of savings.** Investors who bought bonds, life insurance, annuities or similar things were simply cheated without compensation. If we continue to do this, we are going to ruin the United States—and we may drift into dictatorship.

To talk about the industrial power of the United States is just bunk. If the currency doesn't work, the country cannot work. The destiny of the currency is, and will be, the destiny of the nation.

HOW MANY MORE DEVALUATIONS?

PCCE: How many more devaluations of the dollar do you expect in this decade?

PICK: Endless. We may have another devaluation next week already, or maybe only in eight months. The dollar will be wiped out.

PCCE: Many Congressmen have been proposing new import quotas and—

PICK: Forget about Congressmen. They don't know anything. We will lose from these trade barriers.

PCCE: Do you see a danger of trade wars?

PICK: Yes—and we will suffer. The dollar will decline much faster.

PCCE: How bad is inflation in America?

PICK: Very bad. For example, you could buy a good pair of shoes before World War II for about $6. The same pair today is about $35 or $40. In 1940, I bought a suit made to order for $75. Today the same tailor to whom I still go charges $450 for the same suit. I bought my first car in 1948, and it cost $900. The same kind of car today costs about $4,000. I go shopping every two weeks, and my inflation indicator is Pepperidge Farm Bread. Twelve years ago, I paid 16¢ a loaf. Last week, I paid 49¢. You see, the government expropriates everybody who obeys the law.

PCCE: Just how many billions of paper dollars are there in circulation?

PICK: We have now 55 billion paper dollars in circulation. But we also have over 450 billion dollars of government debt in the form of bonds.

PCCE: What's your outlook for inflation?

PICK: I believe this year we are going to have a 15 percent to 25 percent increase in the cost of living. If that happens, we will come close to bankrupting all pension funds. How will they pay? We will bankrupt all institutions of higher learning—they are in trouble already.

PCCE: Would you please reveiw for our readers what happened during America's previous runaway inflations?

PICK: Runaway inflation during the Revolutionary War lasted four years. At the end of that time, the Continental dollar was worth nothing. It expropriated the money of people who believed what the government said. There, you see, none of the government promises can be taken seriously. Governments all over the world have to lie, to lie and to lie to remain in power.

During the Civil War, the dollar depreciated about 60 percent in a few months. There was a bad inflation during World War I. And since 1940, the dollar has been reduced from 100 cents to about 26 cents. Because the dollar was the kingpin of the world's monetary system, the dollar, like venereal disease, infected all other currencies around the world. It's a fantastic story.

A COMMUNIST SOUTH AMERICA?

PCCE: Could you tell us about runaway inflations in other countries?

PICK: In France, during the French Revolution, runaway inflation ruined the economy. During 1924-1926, the French franc depreciated 54 percent against the dollar. Since 1926, the franc has been devalued 19 times.

In China, from 1945 to 1949, runaway inflation completely destroyed the currency—some 425 million Chinese monetary units were equal to one U.S. dollar on the black market. This paved the way for Chairman Mao. There were also runaway inflations in Germany, Hungary, Austria and Russia. They led to destruction and dictatorship.

Argentina, Brazil and Chile are a few countries suffering through runaway inflation now. Brazil, for example, now devalues the cruzeiro about every two months. I was in Brazil a few years ago when they declared 1,000 old cruzeiros equal to one new one. Something like that is going to happen here with the dollar.

Runaway inflation is leading to communism in South America. Bolivia is Communist, Uruguay is Communist, Chile is Communist, Peru is half Communist, Ecuador is half Communist. If Argentina goes Communist, then Brazil falls, and the whole of South America will become Communist. And all of this goes back to the inflationary policy of the United States.

Worldwide, since World War II, there have been over 1,230 devaluations. Soon we will go through the wringer.

PCCE: We've heard rumors, but do not have anything solid on this, that the Treasury Department is printing another currency which could possibly replace the dollar. Do you know anything about it?

PICK: Yes, the Treasury did this last year. The new one dollar bill would remain green, but with a new design. The five dollar bill would be blue, the ten dollar bill beige-brown, the 20 dollar bill red.

Isn't this a tragedy? We have mastered polio, diphtheria and tuberculosis; we have the pill; we've landed on the moon—but we are unable to master currency.

We are vulnerable everywhere. For example, some of the oil princes in the Middle East are increasingly reluctant to accept dollars. Kuwait, Saudi Arabia—these countries buy gold. In the event of further dollar crises, we will see a stoppage in the flow of oil to the United States.* But if we had a gold-convertible currency, nothing could happen to us.

PCCE: What is your outlook for the stock market?

PICK: Panic.

Look—I cannot discuss the stock market because I want to drown all securities analysts in the East River. They are criminals. They cheat. There is not one man who figures a corporation's worth in terms of real money.

PCCE: So even if you have a portfolio of Blue Chip stocks like General Motors—

PICK: There is no such thing as a Blue Chip stock. My torture cabinet is unlimited.

PCCE: What about municipal bonds?

PICK: They stink.

PCCE: We understand the Federal Deposit Insurance Corporation is virtually bankrupt, if it's not already bankrupt—

PICK: I hope so.

PCCE: How would this affect people with savings accounts?

PICK: They will have a lot of trouble. Then no bank account would be safe.

PCCE: How safe would a person's assets be in Switzerland?

PICK: Switzerland has 100 percent banking privacy, and the Swiss have *never* confiscated banking assets. If you had your money in Swiss francs before the February 12 devaluation, you would have something like 18 percent more

*Subsequent events, of course, proved the validity of this May 1973 prediction.—Editor.

money now. Furthermore, if an American has money in Switzerland, he will escape the inheritance tax.

PCCE: Can runaway inflation hit Switzerland?

PICK: No, no, that can't happen—and has never happened in Switzerland. The Swiss franc is backed by gold, so there's a limit on inflation.

PCCE: Can people buy gold in Switzerland?

PICK: Yes, in Switzerland, people are completely free to own gold.

PCCE: Do you recommend buying bags of silver coins for financial protection?

PICK: Yes.

BEST BUYS IN GOLD COINS

PCCE: Do you recommend buying gold coins now?

PICK: Yes, particularly British gold Sovereigns. In the past five years, the price of gold Sovereigns has more than doubled.

U.S. gold Double Eagles, though, have seen a more spectacular price rise—increasing three to four times in the past five years.

Of the two most popular gold coins, British gold Sovereigns are the better buy, carrying a smaller premium over the free market gold bullion price.

PCCE: How high do you believe British gold Sovereigns and U.S. gold Double Eagles may go in the next few years?

PICK: British Sovereigns may go to $70 or $80. Double Eagles, to $550.

6

A Conversation with a Swiss Banker

Hans Weber

Today, with America racked by economic and constitutional crises, many people are transferring a portion of their assets abroad. Although firm statistics are not available, it is currently believed that Switzerland is the primary recipient of these funds—and for good reasons.

For instance, Switzerland has a policy of traditional neutrality going back 458 years. It has been at peace continuously for almost two centuries, even avoiding the two world wars. And to this day, Switzerland remains free of entangling alliances. It does not belong to the European Common Market, the North Atlantic Treaty Organization, the International Monetary Fund or the United Nations.

Another reason is Switzerland's remarkable stability. Switzerland has never had a violent revolution or civil war. It has never suffered through a runaway inflation. The Swiss government has never confiscated banking assets.

Switzerland is also one of the few countries in the world to offer strict banking privacy. Swiss bankers have maintained the confidentiality of client business whether pressured by Nazis, Communists or the tax agents of all nations.

For these reasons, and because it is located in the heart of

Europe, Switzerland has become the banking capital of the world. With this in mind, we decided a consultation with a Swiss banker would be particularly timely, now.

Accordingly, we talked with Hans Weber, managing director of the Foreign Commerce Bank, a bank with which PCCE has had dealings for several years. Mr. Weber studied economics in Zurich and London. After working for American Express, he joined Foreign Commerce Bank where he has served for the past 10 years. With offices in Zurich and Geneva, this is one of the fastest-growing Swiss banks. The Foreign Commerce Bank specializes in offering personalized service to clients in the Unites States and other countries around the world.

This interview took place in PCCE's offices during one of Mr. Weber's visits to New York City, and was published on December 15, 1973. He began by discussing the Swiss banking tradition.

PCCE: Perhaps the most famous aspect of Swiss banking is its privacy. Could you comment on this?

WEBER: It is a crime in Switzerland to disclose any information about an individual's bank account. You can go to prison for violating the bank privacy law.

This policy protects the assets both of Swiss—who value privacy highly—and of people in countries threatened by dictators. The Nazis, for example, attempted to discover which of their citizens had accounts in Switzerland. Those efforts brought on new measures to insure banking privacy.

Swiss banking privacy is not, though, intended to protect criminals. Bank records may be opened up if a foreign government can prove to a Swiss judge that a man is a criminal—that he has done something which is also a crime in Switzerland.

Swiss tax agents may also examine bank records. However, they, too, are subject to criminal penalties for disclosing any information about an individual's account.

Despite these exceptions to the general rule of strict privacy, a person has far more privacy in a Swiss bank than in an American bank.

PCCE: Can U.S. Internal Revenue agents ever gain access to an individual's Swiss bank records?

WEBER: No, because tax evasion is not a crime in Switzerland.

PCCE: Do any other countries have this kind of banking privacy?

WEBER: Yes. At the northern border of Switzerland, the small country of Leichtenstein has it. Americans, though, will have a hard time opening an account there, as the banks are not really set up to handle international business.

Luxembourg also has a bank privacy law. Austria has a good banking tradition. There isn't a bank privacy law in Austria, but it's more or less understood that no information comes out of an Austrian bank.

PCCE: Would you give our readers a quick profile of the Swiss banking industry?

WEBER: We have about 400 banks with eight times as many branches. The three biggest Swiss banks—Swiss Bank Corporation, Union Bank of Switzerland and the Swiss Credit Bank—form one category. They are comparable to the largest American banks. The problem in dealing with one of the big three is that, like large American banks, they're depersonalized.

Another group of Swiss banks specializes in serving the local people. Other banks deal primarily or exclusively with other banks—banker's banks, as we call them.

Then you have international banks that primarily serve clients in other countries. Foreign Commerce Bank is one of these.

We're a medium-size Swiss bank, with assets of about 149 million Swiss francs, or about $49.5 million. I should add that you cannot judge a Swiss bank only by its assets, because Swiss banks also serve as brokers for many clients. Portfolios held or managed for clients are not counted as part of a bank's aasets. Like all banks in Switzerland, the Foreign Commerce Bank is subject to periodic audits by independent auditors.

If I may put in a word for my bank here, we're in the happy

circumstance of being small enough to offer our customers personalized, efficient service, while large enough to satisfy all their banking needs throughout the world.

BANK FAILURES—CAUSES AND CURES

PCCE: What guidelines does your bank follow in its investing?

WEBER: Very conservative. Foreign Commerce Bank considers stocks speculative, and never invests in them for its own account. Foreign Commerce Bank invests its money in Europe and the United States, whereas many American banks are involved in South America and the Far East, where the risks are much greater.

Foreign Commerce Bank does not make any long-term loans. All our loans are secured by easily negotiable collateral, such as stocks, bonds, mortgages, gold and silver. I would say 80 percent of our loans are callable daily.

By contrast, U.S. banks tend to take in short-term deposits, then turn around and go into longterm loans. So if large numbers of depositors want their money, the bank won't be able to recall the loans—and you have a bank failure. This is why many U.S. bank failures happen.

I'm not saying it's impossible for a Swiss bank to fail, but Swiss banks are far less likely to fail than American banks. There have been failures in Switzerland, but practically no depositor has lost any money. The last failure was a branch of the United California bank, in 1970. Even then, no depositor lost a penny.

PCCE: Turning now to currencies, how do you view the Swiss franc and Deutschemark?

WEBER: The Swiss franc is one of the most stable currencies in the world. It's true we're currently having about an eight percent inflation, as every European country has. But the Swiss franc has never been hit by runaway inflation. Swiss francs in circulation are 70-80 percent backed by gold. The Deutschemark is also a strong currency.

PCCE: Do you think it's important that currencies be backed by gold?

WEBER: There are two opinions on this. One says that every currency should be backed by gold. The other says that gold should be completely demonetized. My personal opinion is that the only solution to recurring monetary crises is to back currency with gold. I don't think the Special Drawing Rights [SDR's], or "paper gold," will work.

In particular, if the dollar remains inconvertible, more and more people will question it.

PCCE: Do you view the Swiss franc and Deutschemark as short, intermediate or long-term investments?

WEBER: Something for the long-term. By this I mean many years, looking toward your old age. Short-term speculation is very, very dangerous. Some people have made money, but a lot of people have lost money.

Short-term, I expect the dollar will gain against the Swiss franc and Deutschemark, because there's a dollar shortage in Europe. [Editor: This projection has already proven true.] Nobody has been withholding dollars lately, and there are dollar commitments to meet.

Long-term, though, I would definitely recommend Swiss francs.

PCCE: Zurich is one of the leading, and perhaps the biggest gold market in the world. Could you explain for our readers how this market works?

WEBER: The three biggest Swiss banks form the main gold market, or gold pool. The trading is mainly done by telephone. The price of gold bullion, British gold Sovereigns and U.S. gold Double Eagles is fixed according to the supply and demand situation.

PRINCIPAL FACTORS IN TODAY'S GOLD MARKET

PCCE: What do you think are the main factors in today's gold market?

WEBER: On the demand side, many industries, from jewelry to computers, buy gold.

And gold is used as a hedge by private investors. Looking back at bad inflations, we've seen that in Germany after World War I, things got to the point where it took a billion marks to buy a loaf of bread. But people who purchased gold preserved their capital. Some day, people in many Western countries may have to have gold coins just to buy food for their families.

In the Middle East, much of the oil money is put into gold. If the oil producers continue buying gold at the current rate for the next decade, they'll own something like 50 percent of the world's gold. India, too, is buying gold, though those figures are not available.

On the supply side, South African gold production is down despite higher gold prices. As the price goes higher, it becomes profitable for the mines to work lower grade ores. So the amount of gold produced each year actually declines. In the future, though, if mines expand their operations, gold output may very well increase.

An American cannot hold gold bullion, but as a European, I would hold both gold bullion and gold coins. Gold bullion is the better buy. You could not, though, purchase food or other necessities with a gold bar—you'd have to use coins. So in Switzerland, where people are perfectly free to own any form of gold, there's a lively market for both bullion and coins.

PCCE: What's your outlook for the price of gold?

WEBER: In any market, you'll have technical market corrections, so the price can drop from time to time. The long-term outlook, though, is definitely for higher gold prices. I think investors who want to preserve their capital should include gold in their portfolio.

PCCE: What do you think of silver as an investment?

WEBER: I wouldn't put all my eggs in one basket, so yes, I would recommend silver, too. Here, again, I would divide the funds between bullion and coins. Silver bullion is the most economical way to buy silver, but in a crisis, you'd only be able to buy things with silver coins.

PCCE: Could you suggest a model portfolio for investors interested in preserving their capital?

WEBER: If you have $100,000 to invest, I'd recommend putting about $25,000 in Swiss franc time deposits in Switzerland; $25,000 in gold stocks; $20,000 in gold coins; and $30,000 in silver.

7

The Future for Silver

Charles R. Stahl

Charles Stahl is one of this country's leading experts on silver. Born and educated in Poland and Hungary, Mr. Stahl has lived through runaway inflation, depression, wartime military invasions, and concentration camps at Auschwitz and Mauthausen. Yet Mr. Stahl and his family not only survived these catastrophes—they also preserved their assets by investing in silver, gold and platinum. "Sometimes precious metals are more important than anything else," Mr. Stahl says, "because they give you a certain security nothing else does."

Early in his career, Mr. Stahl was a director of a major European jewelry business with branches in Austria, Poland, Hungary and Switzerland.

Since 1949, Mr. Stahl has been president of the Economic News Agency and publisher of Green's Commodity Market Comments, *a biweekly statistical report and analysis of silver, gold, platinum and currency markets.* Green's *is a leading publication in the commodities field and is published 24 times each year at Box 174, Princeton, New Jersey 08540. A one-year subscription is $120.*

The editors of PCCE's Gold & Silver Newsletter *are pleased to give you this exclusive interview with Charles Stahl. The interview was held in mid-September, 1972, at Mr. Stahl's office in Princeton, New Jersey. Here Telex and Reuters News Service machines bring in the latest news about silver, gold, platinum and the monetary situation. Mr. Stahl's personal research library is perhaps the largest library in America on precious metals.*

Mr. Stahl was happy to discuss one of his favorite subjects in depth—precious metals, especially silver.

PCCE: How do you view the current situation with silver consumption in the United States?

STAHL: This year, silver consumption will increase substantially. In the first half of this year, consumption increased 12 percent. If you apply the same 12 percent to the second half of the year, you'll get a total silver consumption for 1972 of 148 million ounces.

PCCE: What would you estimate silver consumption to be for next year?

STAHL: This is a difficult question, because it depends on how well the economy will be doing. I believe it will be at least 148 million ounces of silver.

Some demands for silver are growing very rapidly now. For instance, the use of silver in medallions, plates and other commemorative items. This is an industry essentially created by one man named Joseph Segel. He started the Franklin Mint. The first year, sales were $1.5 million, and a few hundred thousand ounces of silver were used. In 1971, sales were close to $60 million, and more than seven million ounces of silver were used. This year, the Franklin Mint alone will use somewhere between eight and ten million ounces of silver for its medallions and plates.

PCCE: Do you think this rapid growth of the Franklin Mint can be sustained?

STAHL: No, it cannot be sustained to the same extent. However, last year the private minting industry mushroomed. There are so many mints now that practically every famous American appears on some silver medallion or silver plate. You have the Franklin Mint, the Washington Mint, the Lincoln Mint, the Danbury Mint, many more. At least 20 or 30 mints are operating in the United States.

In my opinion, some of the mints are overestimating their silver consumption. But as far as the Franklin Mint is conerned, I am convinced that, with the excellent merchandising techniques Mr. Segel employs, he will be successful. He might not keep growing at the same clip, but he estimates that sales will be something like $350 million by 1975. Since his sales are based to a great extent on silver, that would bode well for the consumption of silver.

PCCE: How much silver would be consumed if the Franklin Mint reached their sales objectives?

STAHL: Well, I visited the Franklin Mint. It's a beautiful mint, nicer than the government mint. It is the most modern and best-managed mint I've ever seen. Their ratio of gross sales to silver used is seven to one, so if sales were $350 million, 50 million ounces of silver could be used.

PCCE: Do you think silver plates and medallions are good silver investments?

STAHL: No, they're not investments. This is collecting. Collecting may be very profitable sometimes, but that's not the main objective of collecting. You see, a collector will pay for something which he likes, regardless of the intrinsic value of it.

PHOTOGRAPHY

PCCE: Eastman Kodak, America's largest silver consumer, posted record sales last year, and have been setting new records this year. Walter Fallon, Kodak president, estimates that sales of the photographic industry as a whole will increase 15 percent a year this year and in 1973—which would mean increasing industrial silver consumption about ten million ounces. Do you agree with this projection?

STAHL: Yes. Most likely ten million more ounces of silver will be used by the photographic industry.

Now Kodak is marketing a very small, inexpensive camera that is extremely popular. The film is smaller than conventional film, but still silver is being used.

Moreover, I believe vast markets will be opening for the photographic industry. Think about the untapped markets for cameras in Russia, all Asia—people who have never seen, or very rarely, a good, inexpensive camera.

Another point—the cost of silver in film is strictly nominal. The photographic industry has a gross profit on film of about 80 percent. If film is sold at the factory for one dollar, 80¢ of this dollar is profit. So silver could go to almost any price, and it wouldn't make sense to get a substitute.

PCCE: What do you believe will happen to the demand for silver if the price increases sharply?

STAHL: As far as the photographic industry is concerned, the price of silver could go to something like $6 or $8 per ounce, and there would be no need to increase the price of film. Of course, they would increase the price, but there would be no need.

PCCE: Do you expect a substitute material for silver will be developed that could cut back photographic silver consumption?

STAHL: Not soon. There are systems that reproduce with little or no silver—for instance, Xerography.

But you see, silver gives film high speed, so it can be used under almost any lighting conditions. So far, nobody has even come close to finding a good, high speed substitute for silver in photography—and it may take ten to 15 years even to dream about a substitute for silver. I would put it this way: One of these days, somebody should find a good substitute for silver in photography, because there is not that much silver around.

ELECTRONICS

PCCE: Silver conducts electricity better than any other

metal, and its uses in electrical contacts have expanded, so now it can be found in almost every on-off switch and electrical appliance from electric toothbrushes to dishwashers. Do you expect these electronic uses of silver to increase in the months and years ahead?

STAHL: Very much so. For the first time, in the second quarter of this year, the contacts and conductors consumption of silver exceeded the consumption in photography. In the second quarter, consumption in photography was 8.3 million ounces, and in electronics, 8.6 million ounces. Altogether, in the first half of this year, 16.5 million ounces were consumed in contacts and conductors, and 18.2 million ounces were consumed in photography.

PCCE: Do you expect this trend to continue?

STAHL: Yes, because the economy is doing very well. When the economy is doing well, people spend more money. There's greater consumption of silver in photography—though in photography some silver is recovered. In electronics recovery is impossible.

PCCE: Chender Associates, in their massive study of world silver supply and demand, estimates that between 1965 and 1970, the consumption of silver in refrigeration grew 40 percent from ten million ounces to about 14 million ounces. Do you expect this growth to continue?

STAHL: When the standard of living increases around the world, consumption of metals, and particularly such an important precious metal like silver, has to increase.

Refrigeration is a good point. For example, very few people have electrical refrigerators in the Soviet Union, and I'm convinced even fewer people have refrigerators in China. Now the Soviet Union will spend about $55 million to build five silverware factories. If they're going in the direction of doing something for their consumers, then obviously they are going to buy or produce refrigerators—and use more silver for this as well as other products.

Moreover, the Soviet Union used to sell eight million ounces of silver a year in London, but for some time now, none has been sold.

SILVERWARE

PCCE: The president of the Silver Users Association reports higher silver prices have resulted in declining silverware sales. What do you think?

STAHL: He should know better. Total sales of the silverware, hollow ware and silver plated sterling industries were going up as long as the price of silver was also going up. The moment the price of silver began to decline, the sales began to decline. And this is normal. You see, people buy silverware both because of its beauty and its intrinsic value.

OTHER TECHNOLOGICAL USES OF SILVER

PCCE: How rapidly do you believe technological uses of silver will be growing?

STAHL: I am extremely optimistic on that. I would say there are so many avenues and so many new applications of silver that I can see nothing but growth in the consumption of silver.

PCCE: What are some other important technological uses of silver?

STAHL: Silver is used as a dry lubricant. Silver is used as an electrode. Silver is used in a fungicide and a bactericide. Silver is used as a catalyst. Silver is used to purify water in swimming pools.

Silver is used in seeding clouds. It can be done for two purposes. One is to stop or change the direction of a hurricane. The other is to produce rain. Now it is used extensively in the Soviet Union and in France. One ton of silver iodide is used for that purpose, and it has about 14,000 ounces of silver. The ratio is 54 parts of iodide to 46 parts of silver.

Harnessing the sun's energy through the use of silver mirrors is being tested in Europe. In the future, it's possible this will be a source of "clean" energy. These silver mirrors can create heat up to 5,400° F without any pollution.

Now, though, the main uses for silver are in photography, electrical and electronic products, silverware, jewelry, brazing alloys and solders, bearings, catalysts, mirrors, dental and medical supplies, and of course commemorative medallions and plates.

PCCE: What about the new TV cassettes coming onto the market?

STAHL: This is a growing industry that can have great impact. There are something like 90 million TV sets in the United States, and if every set eventually uses five cassettes per week or per month, then the amount of silver required will be fantastic. Better methods to produce silver will have to be found—otherwise, these TV cassettes will create a big upward pressure on the price.

It's only a question of merchandising. Motorola was supposed to be the first company to develop the cassettes and the TV set using them, but it is not yet available to the public. Some large corporations are now using these TV cassettes, as are some educational institutions. It is going in the right direction, but might take a number of years before this new industry comes to a full blossom.

SILVER PRODUCTION

PCCE: Over 65 percent of newly-mined silver in the United States comes as a byproduct of copper, lead and zinc production. How badly do you believe pollution controls for mines and processors will affect silver production?

STAHL: Some smelters in this country have already been closed as a result of pollution controls.

And there's another problem. Pollution or no pollution, there's a question of how much lead will be consumed. If the consumption of lead declines, then the production of silver will decline—nobody's going to mine a lead deposit for the silver content alone. And nobody's going to mine a copper deposit for its silver content either.

In every country but one in the world, silver deposits are near the earth's surface. You can find silver easily. And if none has been found until now, you aren't going to find any.

In the United States, you must go 8,000 to 10,000 feet into the earth to recover silver. It would take plenty of capital—and five to seven years from the moment you decide to start a mine until you are ready to produce. Where are we going to get more silver until this happens?

PCCE: The high cost of starting a new mine now doesn't include the cost of inflation over this period?

STAHL: That is correct.

PCCE: What's been the trend with copper, lead and zinc production?

STAHL: Copper production has usually been increasing, and lead production may decrease, because its consumption in gasoline may be stopped completely.

You see, in this country we use something like 250,000 tons of lead a year as an additive to gasoline, to have a higher octane. Now lead can be replaced completely with platinum. It would cost a pretty penny to the refiners, but this can be done. And since the government insists we cannot have more than 0.03 percent of lead per gallon by 1976, obviously there will be less consumption of lead, and thus less production of silver which comes from those lead-silver deposits.

PCCE: How badly do you think silver production will be affected by the nationalization of Anaconda and other major silver producers in Chile?

STAHL: Very little silver comes from Chile. Chile is not important. The only countries which are important for silver are Mexico, Canada, the United States and Peru. These countries produce about 180 or 190 million ounces out of the worldwide total silver production, about 250 million ounces. In the United States, we produce about 40 million ounces—but we use 140 million ounces, so there's a gap.

PCCE: For more than 30 years, the Sunshine Mine in Idaho was the largest silver producer in the United States. As you know, a tragic fire recently swept through the mine, killing 91 miners and stopping silver production there. In *Green's*

Commodity Market Comments, you suggested it would be a long time before the mine would reopen.

STAHL: When everybody "knew" the mine would reopen in three months, I suggested the time of reopening was unknown. I said it would take at least to the end of the year before we can even think of reopening the mine—and even that is in question.

We had a similar case where the government closed a mine in Pennsylvania, about two and a half years ago. The mine still has not reopened. Similarly, the Sunshine Mine cannot be reopened when the executives desire. The mine can only be opened when the U.S. Bureau of Mines gives its permission. Now the U.S. Bureau of Mines just conducted an investigation; it was completed, but there is no report. This is because inspectors for the U.S. Bureau of Mines were unable to find out what the cause of the fire was. They could not go down to two levels, to the 5,300-foot and the 5,500-foot level where the fire is still burning.

PCCE: The fire is still burning now?

STAHL: Yes. It doesn't mean you cannot reopen a mine. You could. The part of the mine still burning could be isolated, but before you could do that, the U.S. Bureau of Mines would like to know why the fire started, and that can take a while. At the present time, the best estimate by people close to the mine is that they may reopen by this December.

PCCE: How long has the fire been burning?

STAHL: Four and a half months, since May 2.

PCCE: What will be the loss of silver production?

STAHL: About two million ounces of silver per quarter. So by the end of the year, the loss would be about six million ounces. About 40 million ounces of silver are produced each year in this country, so the fire at the Sunshine Mine represents a 15 percent loss.

PCCE: Do you see any new mines developing in this country?

STAHL: No. Not only don't I see any new mines developing, but I see three projects which were closed. One in which something like $6 or $9 million was spent in Idaho by American Smelting and Refining. Another where something like $3 million was spent by Day Mines and other companies. These projects were closed because the companies concluded that at the current price of silver—then $1.30 to $1.50 per ounce—it just didn't make economical sense to proceed.

Of course, when silver goes over $2, these companies will start thinking in terms of reopening. But in the meantime, we don't have the silver production.

PCCE: What's your outlook for silver production in other parts of the world?

STAHL: There's no place in the world where silver production can be increased substantially. What's available is being produced to the hilt. There is some silver in Mexico and other places, but the deposits are known. Old mines are being closed, while few new mines are replacing them. The only new mine discovery we know about is the one Texas Gulf had in Timmins, Ontario. In six years, this mine produced about 55 million ounces of silver—making it the largest silver mine in the world. But the largest amounts of silver were produced in the first, second and third years, because Canadian taxation permits very quick amortization.

Even if the price of silver reaches $4 per ounce, I don't see much more silver production for at least another five years. You will have some increase, because some silver will be mined at $2 per ounce, some at $2.50, but the big bulk of silver that would completely satisfy our domestic demand would need a price of $3.50 to $4 per ounce—in today's dollars.

SILVER SALVAGE

PCCE: In 1970, 49 million ounces of silver were recovered from scrap, which was down about 60 percent from the year before. In 1971, the production of silver from scrap dropped again, to about 46 million ounces. Could you explain this?

STAHL: It had to drop, because the price of silver dropped. Secondary recovery didn't make sense.

Some companies claiming to do well in this ran into financial trouble. For instance, Spiral Metal, Inc., a processor. This company was collecting silver all over the United States. Spiral lost $6 million in 1971 and $6 million so far in 1972. Another company attempting to recover was MacAlpine, Inc., in Rhode Island. It went bankrupt.

Moreover, in previous years, coins were melted, and you had enormous recovery. But nobody melts coins anymore.

I should point out that many of the statistics for secondary silver recovery are unreliable. This is because you don't know what they mean by secondary recovery. For example, at Eastman Kodak, the input is something like 45 million ounces of silver. But the net consumption is about 30 million ounces, because there is *in plant recovery.* Year end consumption figures are net, and the in plant secondary recovery is deducted from it.

I believe the price of silver has to be well over $2 per ounce for secondary recovery of silver to pay off. Otherwise, it's just a waste of time.

PCCE: How much silver can be recovered from Polaroid films?

STAHL: There is a question here. In color films, less silver is used than in black and white. For practical purposes, in conventional color film, you can retrieve as much as 80 percent or 90 percent of the silver. But with Polaroid cameras, that doesn't work—the Polaroid process gives you a positive and negative immediately, and you throw away the negative. Nobody can collect silver that is lost this way, so you have an important loss of silver.

PCCE: Do you expect the secondary silver salvage from photography and other sources could ever help close the production deficit?

STAHL: No, it is impossible.

INDIA'S SILVER

PCCE: How much silver would you estimate came out of India last year?

STAHL: Last year, about 15 million ounces. You don't have to estimate, you know these things. You see, there are very precise statistics on how much silver is being exported from Dubai to England. That is the only thing that is happening to this silver, so you can know to the last gram how much silver is smuggled out of India.

In the first six months of this year, the amount of silver coming out of India dropped dramatically—to a mere 560,000 ounces.

PCCE: How would you explain that?

STAHL: First, the price was too low. At those prices, Indians are buyers, rather than sellers.

Second, India sells silver or gold only in times of distress. In India you don't have banks. People carry their values on their hands and arms—*bangles* of silver. Only when they need money for food do they sell this silver. When there is a bad crop, some silver comes out of India. In 1968, when the price of silver was at the peak of $2.56 per ounce, there was a bad crop in India, and 60 million ounces of silver did come out of India. But then the silver coming out of India dropped to 25 million ounces in 1969, 20 million ounces in 1970, 15 million ounces in 1971, and only 560,000 ounces in the first half of this year.

This decline shows up in the British statistics. It now looks like the silver stocks in England will drop (mostly due to export) this year by 80 to 90 million ounces—an enormous drop. At the peak, England had about 200 million ounces of silver stocks, but by the end of the year they probably will be down to about 80 million ounces. While England doesn't produce any silver, it consumes between 25 and 30 million ounces a year.

PCCE: Will much more silver be coming out of India in the future?

STAHL: You cannot count on the supply. You see, the bulk of silver in India is in religious objects, objects of art, roofs of pagodas, that sort of thing. This silver is not going to be melted for the silver users here.

OTHER ABOVE GROUND SUPPLIES OF SILVER

PCCE: Do you think there are sufficient above ground supplies of silver to satisfy demand in America?

STAHL: No. It is true there are above ground supplies of silver, but are you going to melt your silverware to sell it? Is your wife going to break her bracelet to give it to the users so they can have a cheap silver supply? Certainly not. So the above ground supplies of silver have to be specified. The only one that counts is bullion.

By the end of this year, the worldwide bullion supply should be below 300 million ounces, out of which silver users around the world must carry about 150 million ounces as inventory. Therefore, the floating supply of silver bullion is really very small.

PCCE: There have been substantial withdrawals of silver from the Comex warehouses, and I've heard much of this silver has been transferred to the user's inventory. Do you think this is true?

STAHL: Yes, because the inventory at Comex decreased by 16.8 million ounces, while the users' inventory increased by 8.5 million ounces.

Here I should emphasize that the silver in Comex approved warehouses is the most expensive in the world. First, you have to deliver your silver to one of the five warehouses in New York. Then the silver has to be tested, registered and so on, and it costs you a lot of money. This would be the last silver that anybody would use. That any silver goes out of Comex is the best proof how tight the supply of silver is today.

International Silver Company, a leading producer of silverware, is a good example. Their silver stocks were down to 830,000 ounces according to the most recent report of June 15. In my opinion, these stocks could have lasted about four to six weeks. So International Silver must have withdrawn silver from Comex stocks, where, according to its president, it carried a long position of 2.9 million ounces.

Another example, Engelhard Industries recently withdrew a lot of silver from Comex approved warehouses, because Engelhard's silver stocks dropped dramatically.

Withdrawals like these will begin to exert a steady, upward pressure on the price of silver.

PCCE: What's the current situation with silver imports?

STAHL: So far this year, we imported, net, 22 million ounces of silver. On a yearly scale, that would be 44 million ounces for 1972.

Here's the overall supply-demand situation for the first six months of this year:

The total amount of silver available for consumption consisted of 21 million ounces from new production, 26.4 million ounces from secondary recovery, 16.8 million ounces from Comex stocks, and 22.1 million ounces of net imports, for a total of 86.3 million ounces.

The net consumption was 71.2 million ounces, and the users' stocks increased by 8.3 million ounces for a total of 79.5 million ounces. There is a differential of 6.8 million ounces which is unexplainable, but that happens often with silver statistics.

So this is how the deficit between the demand for silver and new production is being met.

PCCE: Silver soared from $1.29 per ounce to about $1.90, and recently the price dropped off 16¢. When do you think this decline will cease?

STAHL: I believe this break will be of short duration, and before the end of this year, I expect the price of silver will be higher than the $1.95 per ounce that we have seen already.

IS NOW THE TIME TO BUY SILVER?

PCCE: One question our readers will be interested in is, do you think now is a good time to buy silver?

STAHL: Yes. The demand for silver exceeds new production, and that's why I believe there's only one way for the price of silver to go. True, the market can be under some temporary pressure, but in the long run, the price of silver has to go higher.

PCCE: How high do you believe the price of silver will go?

STAHL: I am convinced the price of silver will go to $2.50 per ounce sometime next year. Then I believe the price will stabilize for a while between $2.50 and $3 per ounce. I personally believe the stocks of silver reported in the United States are overstated. And when this fact comes out, it will put a fire under the silver market.

SILVER COINS FOR FINANCIAL PROTECTION

PCCE: What about investing in coins?

STAHL: It is good to have some silver coins, it is good to have some gold coins. You have to insure yourself against the worst that may come. Of course, we hope it won't. It's like taking an insurance policy for your life. You don't believe you are going to die tomorrow. You hope to live as long as possible. The same thing applies to the purchase of silver coins or gold coins.

Assume you have a riot in this country, whether in a certain part of the country or all over the place. You cannot go to the bank, because nobody is going to take your check. Probably the only way to get your food is to exchange your silver coins or gold coins for food or clothing or whatever else you may need. You will only be able to use silver, gold, diamonds and other things with intrinsic value. When you protect yourself against these kinds of disasters, it's irrelevant what the price is.

Genuine silver coins are readily marketable because they can be identified easily. You don't have to go to an expert to check that it's good, bad or indifferent. You look at the quarter coin and see the date 1964 or below. Then you know it's a silver coin, because nobody is going to mint them from something else—it just doesn't pay.

Silver coins are not going to be melted, because they command a premium over bullion. Every time somebody melts silver coins, the ones remaining become more valuable. There are fewer of them, so their numismatic value increases.

Moreover, when you invest in bags of silver coins with $1,000 face value, you have the security that they will *never* go below their face value. Whatever happens to the price of silver bullion, you can always spend your $1,000 of silver coins for $1,000. There is no risk it can go lower. It can only go higher.

8

A Silver Miner Talks About Silver

Philip M. Lindstrom

For 31 years, Philip M. Lindstrom has worked as mining engineer, geologist, mine operator and investment analyst for Hecla Mining Company, one of America's biggest silver producers—with an output of 4.47 million ounces in 1972.

Mr. Lindstrom was drawn into the mining business by the "excitement of finding ore strikes." He started his career as a geologist and mining engineer in Idaho and California gold mines. In 1942, he joined Hecla.

There he mapped rock formations in the course of many exploration efforts. "We met with failure after failure," he says, "because good ore is so hard to find. Many men never find an ore body, but keep on pursuing the geologist's dream. If you find one big ore property in your lifetime, that's quite a victory. Sinking the shaft and getting the ore out isn't nearly so exciting as the discovery."

Most of the time, Mr. Lindstrom served as a mining engineer. "This means," he explains, "planning how to extract metals from the ground as economically as possible. You have to put in a lot of hard work, money, power and machinery. It takes about 100 pounds of 20-ounce per ton ore to produce a single ounce of silver. Taking away the earth's treasures isn't an easy game."

Mr. Lindstrom operated the Radon Uranium Mine, near

Moab, Utah, for Hecla from 1956 until the mine was finally exhausted eight years later. The venture was highly successful—producing over six million pounds of uranium. Profits from the operation enabled Hecla to buy into the Lucky Friday silver-lead-zinc mine in Idaho.

Since 1964, Mr. Lindstrom has headed investment research for Hecla. In this capacity, he investigates mining companies in the United States and Canada to determine their investment potential.

He also writes and lectures widely on silver. His articles have appeared in the Engineering & Mining Journal, American Metal Market, Western Miner *and numerous other publications. He is an active member of the Silver Institute and chairman of the American Mining Congress' Silver Committee.*

During his career with Hecla, Mr. Lindstrom has seen the company pioneer or improve many mining methods and techniques now widely used in the industry. Hecla has grown into a major mining complex. Its silver operations include part interest in the Day Mines and Star-Morning Unit. Hecla also owns a one-third interest in the Sunshine Mine Unit Area, America's largest silver mine, and wholly owns the Lucky Friday.

These primary silver mines are in the heart of the famous Coeur d'Alene district—a silver-lead-zinc "belt" about 25 miles long and four miles wide in northern Idaho. This belt contains most of the few remaining domestic mines that produce silver as a primary product, and is the biggest silver producing district in the world.

Hecla's headquarters are in Wallace, Idaho, a long-established mining community in the Coeur d'Alene. Mr. Lindstrom's office is a research facility containing hundreds of books and periodicals on all aspects of mining. Numerous ore samples are scattered about on his shelves. As we began this interevew, published in January 1974, Mr. Lindstrom relaxed at his desk and was eager to begin talking about silver.

PCCE: The price of silver rose sharply last year, and is now at record highs. How is silver production responding to this, and why?

LINDSTROM: Short-term, primary silver production has dropped. When prices rise, it pays to mine lower grade ore. This extends the life of the mine, but because your hoist capacity remains the same, you cannot remove additional ore from the mine. So silver production has to go down.

Long-term, sustained high silver prices will bring on increased production from primary silver mines.

PCCE: How long does it take for a mine to expand its capacity?

LINDSTROM: When you've got the ore, I would say three to six years. This is to recruit more men, expand the change rooms for the men, install more hoisting capacity, and so on. Suppose, for example, you were sinking a new 4,000-foot shaft in the Coeur d'Alene district. You'd be doing well to make 100 feet per month, and at that rate it would take 40 months for this job alone.

Expansion is very expensive. It would cost about $6 million to sink that 4,000-foot shaft. Then there's equipment—a jumbo drill costs about $50,000; an ore car perhaps $2,500; and a hoist perhaps $1 million. A few years back, we spent $100,000 just on the pipe for pollution control at one mine. It costs about $3,000 per ton of capacity to build a smaller-size mill—for the 750-ton-a-day mill we have at the Lucky Friday, this would mean an outlay of $2.25 million. All these things add up, and it takes time to finance.

So you don't turn these mines on rapidly. There's a lot of lead time. With the energy crisis now, we're finding out this is also true with petroleum and dams.

PCCE: What exploration efforts is Hecla undertaking right now?

LINDSTROM: Overall in 1973, we cut our exploration budget from about $1.5 million annually to $900,000.

We're concentrating our silver exploration efforts in the bottom of working mines. This means the Lucky Friday and Star-Morning Unit.

Exploration is a tedious, step-by-step process. Geological

Hecla's Lucky Friday—

1. Rich silver ore: *4,050 feet underground, this section of a vein may yield 50 to 100 ounces of silver per ton of ore. The dark rock and the glittering, lead-bearing ore contain the silver. The black spots are drill holes used in blasting.*

2. Rock loading: *Blasting fills the "stope," or mining area, with broken ore. Then, as pictured here, an overshot loader fills a mining car. Once loaded, the car is pulled by battery-powered locomotive to the mine shaft where the ore will be hoisted to the surface.*

—Four Views of a Silver Mine

3. Hoist room: *This $1 million hoist installation can lift 750 tons of ore a day from as much as 5,000 feet underground. It also moves men, supplies and equipment into the mine. This picture shows the hoist room at the surface where an operator controls the machinery.*

4. Froth flotation process: *The ore is crushed and ground as fine as flour, then mixed with chemicals in a series of tanks which you see here. The chemicals form a froth with millions of tiny bubbles that stick to the silver-bearing minerals, separating them from barren rock. The result is a cake-like "ore concentrate" sold to smelters who extract the silver.*

mapping is the first step. We may do some surface drilling for samples. If the results look promising, we may sink a shaft and drill underground. We may end up abandoning a site, then going back to it years later, perhaps because our profit situation has improved, or silver prices are higher.

ODDS AGAINST NEW DISCOVERIES

PCCE: What do you estimate is the likelihood of finding a new silver deposit?

LINDSTROM: Well, we don't specifically look for silver. Rather, we look for a good ore body of any kind—whether silver, gold, copper, lead or zinc. The chances of discovering a good ore body are about one in 1,000. That is, for every 1,000 sites where you do drilling or other exploration work, only one is likely to become a producer of metallic ore.

One man recently reported in the *Northern Miner* that in Canada the chances of finding a metal mine are about one in 10,000.

You see, an ore body is a freak of nature. There aren't many available. Of course, when you hit a good ore body, you've really got something. That's the prize we're after.

As for finding silver—most of the primary silver deposits occurred at shallow depths. So they were easy for miners to find, and were worked out long ago. For instance, there was the silver in Laurium, Greece, that Alexander the Great used to finance his wars. The silver mines worked by the Incas in Peru and the Aztecs in Mexico—treasure that sparked colonization of the Americas. Nevada's Comstock Lode, which helped the North finance the Civil War.

Today, most primary silver mines still producing are deep. The Sunshine Mine, for instance, is working ore bodies some 6,000 feet below the surface. But since these are "blind" deposits—with little indication of silver at the surface—they're very difficult to find.

PCCE: How much time does it take from the point when a decision is made to develop a property, until a mine actually begins producing?

LINDSTROM: Generally speaking, I would say four to six years, but it varies widely. The Radon Uranium Mine took about a year to get going. At the Lakeshore Property copper deposit we're developing in Arizona, we started sinking the shafts in December 1969, and won't go into production until early 1975—so, a little over five years.

A TOUGH LABOR MARKET

PCCE: Earlier you mentioned that one of the reasons that it's difficult to expand is the time required to recruit more men. I take it then, that you face a tough labor market?

LINDSTROM: Yes, it's always hard to find good miners—I hold them in very high regard. We have to find men with different skills and different styles of working to handle the various jobs.

A shaft miner, for instance, drives major vertical passages downward. A raise miner drives smaller vertical passages upward. A drift miner drives horizontal tunnels. A stope miner extracts ore from the developed ore body. Each of these jobs requires the use of different equipment, hand tools, ground support techniques, and mining methods.

Miners can make pretty good money. Some prefer to be paid on a day rate "from portal to portal," which means from the time they go down into the mine until the time they return to the surface. The day rate ranges from $31 to $35.

Other miners work at "gypo" rates and are paid according to the number of cubic feet of good ore they mine. Efficient gypo miners, or "contract" miners as they're also called, can make from one and a half to two times more than miners on a day rate.

PCCE: How hazardous is silver mining?

LINDSTROM: Well, I've worked down in a mine for years, and I don't object to my oldest son working down there now.

Mining is probably about as safe as driving a car. The risks are there, of course, but we take extensive precautions to avoid accidents. During the course of a year, many mines operate without a lost-time accident—this is when something happens on one shift that prevents a man from returning to work the next shift.

I think safety largely depends on the man. If he's careless and doesn't respect the hazards around him, he'll be prone to accidents. If he's careful, he'll tend to have few accidents. It's like a lineman on a power line. He's working right with that high voltage all the time. Linemen, though, follow essential precautions and have fine records.

PCCE: How high is your labor turnover?

LINDSTROM: In a long-established mine like the Star-Morning, it's low. Most of the men have their homes in the area, and they don't move around much. Company benefits that increase with length of service—like pensions and vacations—help make things more stable than when I first started in the business.

But in the beginning of a new operation, such as our Lakeshore Property, you may have a comparatively high turnover. It can take a while for men to settle down and get into a routine they like.

PCCE: To what extent are strikes a problem?

LINDSTROM: Well, the Sunshine had a costly strike early in 1973, and the eight-month copper industry strike in 1967-1968 was a bad one. So strikes can hurt in mining just like in any other industry.

We've been fortunate, though, at Hecla to have relatively little trouble. We try to run the mines in a way that will earn the confidence of the men.

PCCE: What caused the tragic fire at the Sunshine Mine?

LINDSTROM: The fire was burning for so long that no one has been able to reconstruct what caused it. It's like a hotel fire—the evidence goes up in smoke.

Personally, I think the company's safety practices were sound. I very much doubt that negligence was behind the fire.

PCCE: You mentioned earlier that you cut exploration—yet prices were generally rising—

LINDSTROM: Price controls: At Hecla, price controls reduced our income more than anything else. Among other things, this means we don't have the money for exploration that we used to have.

Generally, controls are disastrous for marginal producers. In zinc, controls led to the closing of fully half the U.S. refiners over the past three years. Meanwhile, new zinc refineries are being built in Japan, Canada and Europe.

You see, mining companies lost money year after year with the hope that metal prices would go up, and we'd make back what we lost. But then the Cost of Living Council came along and clamped a ceiling on metal prices. These people talked about slowing down inflation, but they didn't know what they were doing. Inflation continued, and many mines were getting strangled by price controls. The Russians couldn't have done a better job on us than the Cost of Living Council did.

When controls force mine operators to cut back production, the ore that's abandoned will stay underground, never to be recovered. So the controls do permanent damage. It's a good thing they were abolished on silver, lead and zinc.

PCCE: How badly do inflation and taxes affect silver production?

LINDSTROM: When you compare our income with the purchasing power of the dollar, you can see we've got our backs to the wall. The only bright spot is the miner's dream that the value of metals will come to be recognized—as is happening with silver over $3 per ounce today.

As far as taxes go, if the government allows mines to retain enough money to replace depleting assets, that's helpful. But when the government imposes a tax on the gross value of mine production, this is destructive. It cuts profits and means that mining in certain veins must be stopped. The result—more ore is left underground.

PCCE: What's the average life of a mine?

LINDSTROM: It varies. Some mines last five years. The Radon Uranium Mine lasted nine years. We closed down the

Mayflower silver mine in 1972, but it had been running for over 30 years. The Star-Morning Unit has been producing lead, zinc and silver for 85 years. The Sunshine and Lucky Friday, for more than 40 years each.

A PROJECTION FOR THE FUTURE

PCCE: At what pace do you expect silver production to grow in the years ahead?

LINDSTROM: It's a great mystery trying to tell what the future holds. General projections I see for expanded production have fallen short for one reason or another. A strike, fire, inflation or something comes along.

I think if silver production increases two percent per year at current prices, we're doing pretty well.

PCCE: Since reaching a low of 45 million ounces a couple of months ago, Comex silver stocks have gone up to over 60 million ounces. What's happening here?

LINDSTROM: I've heard reports that certain Texas oil interests are buying silver in a big way as an investment. The short sellers have to cover their positions by bringing silver in from England and Mexico. The Texas oil interests are taking delivery and storing the silver in Comex warehouses. So while the level of Comex stocks has increased, not all of this is available to supply industrial users, at the present price.

I expect the squeeze on short sellers to continue, and I think we'll see the level of Comex stocks increasing, but no one knows for sure. The supply situation for industrial users won't change much, though.

PCCE: Do you mean that industrial users will continue to have relatively easy access to silver supplies?

LINDSTROM: Yes. I've talked with people at Kodak, Polaroid and other companies that use silver, and they're not worried about the situation. They expect to obtain all the silver they need—at a price.

PCCE: What's your outlook for silver consumption?

LINDSTROM: I'd say the prospects are most encouraging. In part, this is because silver can function in so many ways. And in many cases, silver is the least expensive material that can be used.

In photography, with more and more people around the world becoming affluent, more snapshots are being taken. You've got to have X-rays for health care, so silver is being used for X-rays like there is no tomorrow. Photo-offset is a high quality, fast and economical means of reproduction so here, too, silver consumption is on the rise.

Recently, some eager beaver at an Eastman Kodak lab worked out a way to hook a movie camera to a computer so that data could be visually stored. Film is what Kodak sells, so they're excited about it.

In electronics, silver is being used because it's the best conductor of electricity and heat, doesn't deteriorate, and is cheaper than other suitable materials. Even the silver sulfide that forms tarnish is an excellent conductor.

Silver also kills bacteria and fungi, yet is completely non-toxic for human beings. Silver is used to prevent blindness in babies. It's also used to purify water. At a recent American Mining Congress meeting, we mixed drinks with water from a swimming pool purified with silver. It was a dramatic demonstration of the point, I think.

Photography, electronics, medicine, sanitation, batteries, bearings, brazing alloys, mirrors, coins, ingots and medallions—silver has so many uses.

And of course, many people like to put away some silver as a hedge against inflation. People have been doing this since before Biblical times.

Taking all these things into consideration, I think that rising prices with volatility will continue to be the story in silver.

9

Is Silver Essential in Photography?

Wallace Hanson

The rapid growth of the photography industry—the largest single consumer of silver—is an important factor behind the current bull market in silver. A substitute for silver in photography, if one were to be developed, would be a significant depressant to the market. We've heard persistent rumors of such a substitute for many years, and thought you would be as interested as we in determining whether silver is essential in photography.

So we turned to Wallace Hanson, contributing editor of Popular Photography, *and commissioned him to research and write an article in answer to that question, which we published in December 1973.*

Every roll of film, every snapshot and every movie requires silver. Without silver, photography as we know it wouldn't exist.

To be sure, there are image-forming systems other than silver photography. We may hesitate to call them photography, but only because of linguistic practice and custom. Any means of recording a lasting image by the action of light deserves to be called photography. Though nonsilver systems are called "unconventional" photography, it is silver photography which is the Johnny-come-lately. One of the unconventional means of making pictures actually preceded silver photography.

EARLY PHOTOGRAPHIC PROCESSES

Joseph Nicéphore Niépce (1765-1833), the inventor of photography, first tried a silver process, but turned to other processes when he failed to make a silver image permanent.

He made his first successful, lasting image on an asphaltum or bitumen plate, probably in 1822. Niépce found he could inure it to solvents such as oil of lavender by exposing it to light.

As early as 1826, Niépce made a photograph with a camera by exposing a glass plate coating with bitumen to a sunlighted scene for 12 hours. Most of the inventor's early pictures are lost, but we can be sure they did not clearly delineate shadows—because during the course of a day, the sun moves a full 180 degrees.

To get speed, Niépce, his son and their well-known partner of latter days, Louis J. M. Daguerre, used silver. Actually, the daguerreotype was a solid plate of silver; and this was too profligate a use of the precious metal, so the daguerreotype was replaced by methods that were stingier in their use of silver.

These silver systems were ideal for portraiture, because they did not require the subject to spend a day in *rigor mortis* under blazing sunlight for a likeness to register. So silver photography caught the public fancy. But the unconventional, nonsilver processes were by no means forgotten—they were merely pushed aside to industrial and commercial applications.

In these applications, "unconventional" processes long ago surpassed silver photography in the sheer number of images made. The average secretary makes more "photographs" of business correspondence on a Xerox machine than does the guy who photographs her wedding.

ELECTROPHOTOGRAPHY

In the electrophotographic processes derived from the late Chester Carlson's patents—Xerox copy processes and its licensees—there is an electrostatic charge on the image-forming surface. This electrostatic charge has certain properties similar to those of a bar magnet. Because of these properties, the revealed image tends to concentrate at outlines or edges. This is fine for copying business correspondence—typewritten characters are entirely outlines and edges. But when it comes to photographing either large, solid areas or smooth, gradual tones such as you have in a portrait, electrophotography won't work. It reduces most of the image to outline.

This effect may even enhance the copies of reading matter, so the peculiar effect seen in the Xerox copy is fine for its purpose. None of the electrophotographic processes so far demonstrated have been able to make a continuous-tone print comparable in quality to silver photography.

Many attempts have been made to eliminate the edge effect. For instance, an electromagnetic bar to manipulate the magnetic field of the electrostatic charge will help. Using white dot screens to break up the image into small segments also helps. Each generation of Xerox machine displays improved results.

But even the color Xerox machine, though it will reproduce a lovely color image of lines or halftones, will not make a snapshot that compares with the cheapest Instamatic or Polacolor print.

Samuel B. McFarlane, Jr., of Electroprint, Inc., who licensed his patent to Savin Business Machines Corporation, developed an electrophotographic system using a charged screen rather than a solid surface for his image-forming surface. Once an ink image is formed on the screen, it is "shot through the air onto plain paper without contact or pressure." The screen may break up the edge effect, and being shot through the air may diffuse the ink so that a smooth tone is achieved. Independent witnesses to demonstrations have verified that prints made with the process have a graduated tone.

In another process, called electrophoresis, static charges are not used. Instead, a "memory effect" is exploited. By doing away with the static charge, the edge effect is avoided. The 3M Company, which pioneered electrophoretic photography, has generated a number of products based on it, but they have not been an important factor in photography. The art is there, though, and it exists as a technically feasible alternative to silver photography in color and black and white. This process could be used in a darkroom, but not a camera, because it is too slow and requires up to three electroplating baths immediately after exposure.

But even if Xerography and other electrophotographic processes could produce a continuous-tone image to match the quality of an Instamatic or Polaroid snapshot, they would still have a long way to go. To see why, let's compare the Xerox process with silver photography.

Silver photography is "fast" or "sensitive" to light. Speed and sensitivity may be contradictory traits in people, but in photography, both have much the same meaning. They have to do with how little light you can use to make a good image.

Xerography and other electrophotographic processes take surprisingly little light. Xerography then amplifies the light energy with electrical power, with a high voltage static charge on light-sensitive surface. Just how much amplification goes on in a process like Xerox's is a matter of debate at scientific meetings. One estimate, though, is that the multiplying effect of electrophotography over a system which works by the effect of light exposure alone is 50,000 times.

This is an impressive figure. It vastly outclasses almost all competitive systems. It is key to the success of the modern office copier. Among other things, this massive light amplification permits the use of a lens. With a lens, originals can be copied with the light reflected from them, rather than through them.

Impressive though the figure for electrophotography may

be, silver photography is a good deal more impressive. Through chemical development, *silver photography can amplify the original effect some 100 million times—2,000-fold greater than Xerography.*

Thus, we see there is something very basic involved here. Even if it were practical to make an electrophoto system like Xerox's that could be used in a camera, stupendous improvements would have to be made in sensitivity. Otherwise, the process just would not work for the ordinary person. Simple cameras with their small lenses and snapshot shutters would not admit enough light to create the ghost of an image.

Sensitivity alone, according to B. H. Carroll of Rochester Institute of Technology, makes it "clear why the silver processes are difficult to replace."

But there are still more reasons why silver photography will be around for quite a while. The initial stage of image formation in silver photography is accomplished without an external source of energy. The amplification step, chemical development, is subsequent. Unlike electrophotography, it may be postponed almost indefinitely, at the user's convenience. It can also be miniaturized and hastened as in the brilliant example of the Polaroid Land system.

This freedom from an external source of power or a cumbersome developing mechanism is no trivial advantage. Just imagine taking the likes of a Xerox 914 office copier, and the special electrical cables for it, along on a picnic. Imagine trying to maneuver it at a wedding dance.

When electrophotography is applied to printmaking, such objections are not nearly so weighty. Weak sensitivity can be compensated for by shining more light through the negative or making the exposure longer. True, demands like these have put electrophotography at a competitive disadvantage with silver photography, and the disadvantage has precipitated the commercial failure of some applications.

TELEVISION

Television is the only image-forming system comparable

in speed to silver photography. Indeed, some electronic systems akin to commercial television are more sensitive than silver photography.

The "sniperscope" and police surveillance devices are examples. These use a vidicon image intensifier which GAF's H. F. Nitka says is "at least 100 times faster than fast photographic silver-halide film, but sacrifices considerable detail rendition for this speed." He estimated that the detail of the video system is on the order of one-tenth that of silver photography.

Without some major breakthroughs, television won't match the quality we now take for granted in our snapshots and projected images, and the compactness represented by the pocket Instamatic.

Moreover, television cannot produce a print, or snapshot. Commercial equipment is not now available, and it would be prohibitively expensive to produce. So television doesn't stand a chance of matching the trifling cost of mass-produced cameras.

It is interesting to note that when it became clear television would replace movies as America's mass entertainment, many people in the movie business were certain that the motion picture industry was doomed. Yet, just the opposite came true. Film sales for movies made for television exceed the film sales for theatrical motion pictures in the days of Hollywood's lustiest profits.

CHEMOPHOTOGRAPHY

Are there other photographic processes besides those based on silver or electrical energy? Yes, but the trouble with most of those known is that they don't permit an amplification step, and are impossibly slow. That doesn't mean they are useless—only that they are useless for cameras.

One such process that has been made available in film formats is best-known by the name of the company which first produced it commercially—Kalvar. It and products like it made by Xidex Corporation are classified as vesicular processes. Vesicules are little bubbles. The image consists of little bubbles of air in an "emulsion," a plastic akin to Saran wrap. By ordinary light, the bubbles look white, but in a microfilm projector, the bubbles scatter the light. So where they are present, a dark image is projected.

The vesicular process is capable of high resolution (delineating up to 500 lines per millimeter) and is one of the most permanent types of photographic images, if excessive heat is avoided, and reasonable care exercised. But the Kalvar process is *so insensitive that it can be handled in ordinary room light.*

By contrast, the resolution of the silver image has never been satisfactorily measured, because the measuring grids only go up to 2,000 lines per millimeter—and a correctly processed silver microfilm image is thought to have staying power of indeterminable duration. Silver photography outperforms vesicular process in almost every respect but cost.

The vesicular process is useful for making conversion negatives from slides. But its main application to date is in microfilm duplication.

Another chemical process is the "whiteprint" or diazo process that most engineering drawings are printed on these days. Diazo is a dry process, which is based on light-sensitive iron salts. Both diazo and blueprint processes are low cost, but neither is much of a threat to silver photography, because neither system is as permanent. And both are so slow they require special light sources at close range to expose them.

OTHER PROCESSES

There are numerous other processes. One uses heat, but there are no signs it is being developed into a marketable product. Another process works in ultraviolet light. Still another uses platinum and is therefore quite expensive.

One of the more hopeful processes is called "free-radical photopolymerization." In lab tests it has amplified light about 50 times. It would have to do a thousand times better even to catch up with electrophotography, and much more than that to come near silver photography.

A SUBSTITUTE FOR SILVER LIGHT YEARS AWAY

One could go on and on, but there's no need. The kinds of silverless photographic processes must number in the hundreds. Many are very useful, but are just too slow for most photographers to use. There is no way to guess whether a process will one day replace silver photography. All we can say for sure is that all known processes are very far from doing so. The possibility exists that none ever will.

10

The Future for Gold

Paul C. Henshaw

Dr. Paul C. Henshaw is president, director and chief executive officer of Homestake Mining Company, which operates the largest gold mine in the Western Hemisphere.

Dr. Henshaw first became interested in geology and mining during his college days. "I have a scientific turn of mind," he explains, "but I'm really not geared for the high, pure air which the physicist breathes. I wanted something more down to earth, and found it in geology. For a vigorous young man, it's a vigorous way to spend your life." After graduating from Harvard, Dr. Henshaw earned his M.A. and Ph.D. in geology at the California Institute of Technology.

Then from 1940 to 1943, he served as head geologist with a silver-copper-lead-zinc mine in Peru, operated by Cerro Corporation. He learned how to map rock formations so miners could sink shafts and drive tunnels with optimum access to the ore.

But he was determined to work in the gold mining business. He recalls that "during the Great Depression, I could see the economics professors were walking around with patches on their coat sleeves, while geologists and other gold mining people were riding to work in fine cars. I figured that depressions were cyclical, and that sometime in the future we would see bad times again. As you know, the gold business is a good place to be today."

Over the next decade, he worked in gold mines with the Consorcio Minero del Peru in Peru; in silver-lead-zinc with Day Mines in Idaho's Coeur d'Alene district; and in gold-silver with San Luis Mining Company in Mexico. He taught geology at the University of Idaho.

Finally, in 1953, Dr. Henshaw came to Homestake as chief geologist. He was hired by Dr. Donald McLaughlin, president of Homestake, who had taught him geology at Harvard. Dr. McLaughlin says, "I knew Paul had outstanding knowledge of geology. He also demonstrated the ability to get things done well under pressure."

He was quickly sent to Utah. In the following six years, he conducted wide-ranging exploration for uranium. He learned many techniques that helped improve the effectiveness of Homestake's overall exploration efforts.

Dr. Henshaw's administrative abilities grew increasingly apparent. He became vice president in 1961 and president in 1970. During his 12 years as an executive, Dr. Henshaw has played a key role expanding Homestake into uranium, lead, zinc and silver. These operations helped tide Homestake over the difficult period when inflation was driving costs up, and gold remained fixed at $35 per ounce. They account for Homestake's diversified strength today.

Reflecting on the various jobs he handles as president, Dr. Henshaw says, "the exploration end of the business is still what I enjoy most. That's what I know best, and it's the very lifeblood of a mining company."

The Homestake gold mine in Lead (pronounced "Leed"), South Dakota, though, continues to be Homestake's most valuable property. The history of this mine goes back to the gold rush of 1876, when thousands of prospectors swarmed to the Black Hills of the new Dakota Territory. Two prospectors from Oregon—Fred and Moses Manuel—discovered the underground deposit that came to be known as the Homestake.

In 1877, they sold their property for $70,000 to a syndicate headed by George Hearst, a successful mine operator during the California gold rush of 1849 and the Comstock gold and silver bonanza of 1859. He put together the primary claims for Homestake Mining Company. This provided much of the Hearst family fortune—which his famous son, William Randolph Hearst, was to use in newspaper publishing and politics.

Today, Homestake is the only gold mine still producing in the Black Hills, but it is a huge operation. Ore mining is conducted on 35 levels, from 1,700 feet below the surface to 7,200 feet. Including all levels, there are some 200 miles of workings—an area so vast that travel is only feasible on a major transportation system akin to a city subway.

Homestake handles the entire production process from mining to refining. In 1972, Homestake processed 1.47 million tons of ore, from which 407,397 ounces of gold and 99,319 ounces of silver were recovered.

Homestake's headquarters are in San Francisco. They afford a splendid view of San Francisco Bay, Nob Hill, Telegraph Hill and other landmarks of that scenic city. Dr. Henshaw's office features detailed maps of Homestake's gold mine, and a large chunk of rich Homestake ore speckled with gold. After making us comfortable, Dr. Henshaw began to share his knowledge of gold, which we published in February 1974.

PCCE: Recently, the European Economic Community proposed using the free market gold price in settling accounts among member countries. Could you comment on this?

HENSHAW: For a while monetary officials pretended pieces of paper called SDR's—Special Drawing Rights—could be used to settle international accounts. But they've abandoned that idea and returned to gold. By using the free market gold price, the Europeans have put a new high floor under gold. What those of us who believe in gold thought was inevitable has now happened. I think it will mean further upward pressure on the price of gold.

PCCE: What's your view of the overall economic situation today?

HENSHAW: Governments around the world are falling deeper into debt and printing huge quantities of paper money. As a consequence, people everywhere are squeezed by inflation.

In America, I think it's politically untenable to balance the budget. As long as most people seem to want Uncle Sam to assume a bigger role in our lives, deficits and inflation will continue. We'll be fortunate to escape a massive inflation.

The Mideast war and the Arab decision to raise oil prices sharply only made a bad situation worse. Now every economy that uses large amounts of oil may be thrown out of balance. We're beginning to see this here, now that many factories, airlines and gas stations are laying people off.

PCCE: As, of course, you know, popular interest in gold has been skyrocketing. Do you expect this to continue?

HENSHAW: Yes. As you may gather, I'm very pessimistic about the economic situation. I think the flight into gold has just begun.

And I think this will continue for quite a while. With inflation and the oil crisis worsening, and dishonesty so commonplace in government, increasing numbers of people are turning to gold for protection. Gold has been used as an international standard for thousands of years. Africans in the jungle value gold. Eskimos in the Arctic value gold. Arabs in the desert value gold.

Those who lived through runaway inflation are among the strongest believers in gold. In France, Germany, Austria, Hungary, Greece, India, Indonesia, China, Chile—wherever people have been hard-hit by inflation and the crises it brings on, they hoard gold.

PCCE: To what extent do you think legalization of gold ownership in America will increase the demand for gold?

HENSHAW: I think a lot of people will buy a few hundred dollars' worth. Suppose a half a million or a million people buy only that much gold—it will take all of U.S. gold production for a couple of years. There's no telling how much legalization will add to the demand for gold, but I'm confident it will grow.

PCCE: What do you think the trend will be for commercial gold consumption in the years ahead?

HENSHAW: Worldwide commercial consumption of gold exceeds mine production, and I expect it to continue growing—because gold is such a versatile metal.

The electronics industry is using a lot of gold because it's an excellent conductor of electricity. Gold contact points need make only a slight contact with each other to complete a circuit. Since gold never corrodes, gold contact points and printed circuits provide the ultimate in dependability. Computers and everyday appliances use gold. The San Francisco telephone exchange which you can see from my window is loaded with gold.

Then there's the gold plating industry. A multitude of things are plated with gold. For instance, goblets, candle sticks and dishes. You see gold lettering on everything from peanut butter jars to books.

Some gold is used in medicine, primarily to make fillings for teeth. Gold is also used to treat rheumatoid arthritis and certain kinds of cancer.

An extremely thin film of gold can reflect intense heat, so it's used in jet engines and rockets as lightweight insulation. Gold is used to protect astronauts from dangerous radiation in space. In hot climates, gold on walls and windows of a building can cut your air conditioning bill.

Gold is a uniquely rich yellow—and it stays beautiful forever. Gold is easily melted, easily cast and worked. So the main commercial use for gold is in jewelry—wedding rings, cufflinks, bracelets, pins, watches, you name it.

I think a good deal of the growth in jewelry sales, though, is actually hoarding. For example, I know of one company making simple gold bracelets that weigh exactly one ounce. They're selling like hotcakes—probably for the same reasons there's such a phenomenal boom in gold coins and gold shares. People who never thought of buying gold before are snapping up almost anything made of gold.

DECREASING GOLD PRODUCTION

PCCE: Gold production has been dropping in the face of record high gold prices. Would you explain why to our readers?

HENSHAW: Well, we always try to work as near as we can to our hoist and mill capacity—about 6,600 tons of ore per day. With higher gold prices, it's become profitable for us to mine lower grade ore. So while the amount of ore we process remains about the same, actual gold output is lower. For instance, during 1972, our gold output declined 20 percent from 1971. Yet our income and profits are considerably greater than a few years ago. The situation is similar for gold mines in South Africa.

Over the years, I expect gold production will continue to drop—the faster the price of gold rises, the faster the drop. If gold were at $300 per ounce, I think gold production would be something like a half or third what it is now.

PCCE: As the price of gold rises, can you go back to old mining areas and dig out ore that was previously too low in grade?

HENSHAW: Generally no, because it doesn't pay. There are tremendous rock pressures underground that become greater the deeper you go. To support the rock and reduce the danger of cave-ins, we fill worked-out areas with sand plus a little cement to make it stick together. When we're all finished mining an area, it's completely filled in. It would be prohibitively expensive to dig out all that partially-cemented sand for the ore we left behind.

When mining areas haven't been filled in yet, though, we can go back and mine more ore. We're doing some of this now.

PCCE: How has the labor situation at Homestake changed since the strike in 1972?

HENSHAW: Things are considerably better now largely because the price of gold is so much higher. This has enabled us to pay wages that just weren't possible until gold moved well above $35 per ounce.

We put as much work as possible on a "contract" basis. This is where the pay increases with the amount of good ore mined. Our good contract miners make $15,000, $17,000 and sometimes more a year—which is as well as miners do anywhere in the country.

As a result, our turnover is, I'd say, less than 10 percent. Most of our people spend their whole lives with us. The mine has been going since 1877, so a stable community has grown up around the mine. Miners buy homes in the area and usually retire there. As a result, we have something of a problem providing housing for new people.

PCCE: To what extent are higher gold prices stimulating new exploration for gold?

HENSHAW: Well, many companies are now looking for gold—not just the few ongoing mines that have always done exploration work. At Homestake, we're spending over $2 million annually on exploration. This is up substantially from the past few years.

We're looking at something like 80 to 100 sites a year. We're also stepping up exploration in the bottom of our gold mine. We're down to 7,200 feet and plan to go down to 8,000 feet to develop the extension of the gold ore in depth.

Besides price, the principal factors affecting the level of a company's exploration efforts are the size of its bankroll and its overall company position. For instance, a company with big ore deposits and money rolling in may aim primarily to make current operations more profitable. On the other hand, a small company may put a disproportionately large amount of money into exploration—so that hopefully it will discover a big deposit and grow accordingly.

I should point out that exploration takes a great deal of time. The first step is to describe rock formations at the surface. We'll send teams of geologists and surveyors to make detailed geological maps of a region. We cover practically every square inch of the ground. On the basis of the surface maps, we estimate the likelihood of mineralization underground.

If indications are favorable, we'll drill for ore samples.

Sometimes, when a potential ore deposit is too deep, or below a layer of rock that's difficult to drill, we may dig underground tunnels. These will be about eight feet high by seven feet wide—a major job—to provide room for the machines we'll use in drilling more ore samples.

Analyzing the ore samples enables us to determine rock types, the shape of a vein, its thickness and ore grade. Then geologists make prediction maps to show the expected shapes of geologic structures.

Another technique we use both on the surface and underground is the geophysical survey. This involves transmitting low frequency electrical current into rock, then measuring the flow of current through the rock at a series of points. Since minerals tend to be good conductors, they'll make distinctive marks in the flow patterns. Another technique measures the magnetic field within the rock to aid in geologic study and interpretation of the rock structure and possible mineral content.

Altogether, it may take us two or three years of exploration before we can decide whether bringing a mine into production would be worthwhile.

THE ODDS AGAINST NEW DISCOVERIES

PCCE: What kind of results do you expect from your exploration program?

HENSHAW: To find any kind of metal, I would say this—for every 1,000 exploration sites we look at, perhaps one may become a profitable producer.

Ira B. Joralemon, one of Homestake's directors and a great geologist, put it another way: Out of every seven sites you go out and look at, you'll actually do a little work on one. Out of every seven you do a little work on, you'll spend a substantial sum of money on one. Out of every seven of those, one will become worth something—either to be developed by you or sold to someone else. For every seven of those, one will become a modest producer, while for every seven modest producers, you'll have one very profitable mine.

The odds against finding gold are still greater—partially, of course, because gold is rare. Even in our Homestake gold mine, it takes almost 8,000 pounds of ore to yield an ounce of gold.

Another factor is that surface deposits of gold were easy to find and so were exhausted quickly. I'm thinking of the gold found in streams and rivers that led to the famous gold rushes. The California gold rush of 1849. The rush to British Columbia and Australia in 1859, to Pikes Peak, Colorado and Virginia City, Nevada in 1859, to South Dakota's Black Hills in 1876, and to Alaska's Klondike in 1897.

Most of the world's great gold mines today are deep—in South Africa, going down more than two miles into the earth. This is why the search for gold requires a lot of money, equipment, expert geologists and mining engineers, just to make that one-in-a-thousand shot.

PCCE: How many years of reserves does Homestake have?

HENSHAW: At any particular time, we have from five to eight years of reserves. So you multiply current production by this, and you have our proven reserves—about 7.2 million tons of ore now.

Of course, that's not all. Since exploration and development are so expensive, we aim to find new reserves at the same pace that we can mine them. It doesn't pay to tie up your money in reserves that won't yield any income for, say, 15 years.

Our most important reserves are at our gold mine. This is far from completely explored. I think it will take years and years, plus tremendous investment, to ever mine all the gold out.

And as the price of gold rises, our reserves go up, too. This is because ore that was previously below grade becomes worth mining. Gold output may continue heading down, but the life of the mine is extended.

PCCE: About how much does it cost to bring a new mine into production?

HENSHAW: That's a broad question. If you're talking about an open pit mine where all you do is remove ore from the surface, it may cost only a few million dollars. But it may cost between $50 million and $200 million to start up a deep gold mine like Homestake or the mines in South Africa.

On these major undertakings, it may take you five years just to bring the mine into production, and as much as 12 years to make your money back. So you'd need substantial financing. This would take a lot of time to arrange. The Mexicans have a saying that it takes a mine to make a mine—in other words, the days are long gone when prospectors could make do with only a pick, pan and shovel.

Consider a few of the things that go into our gold mine. We have about 100 miles of track in active use. We use more than 1,100 ore cars and locomotives down in the mine. Our drill-sharpening shop sharpens over 325,000 tungsten-carbide bits a year. In our mill, 750,000 pounds of steel rods and 2.5 million pounds of steel balls crush the ore. Energy consumption for the entire operation is more than 120 million kilowatt hours annually.

Just providing adequate ventilation is a big job. When you're 6,800 feet below the surface, for instance, the rock temperature is 124 degrees. Because water is used in the mining process, the potential humidity is close to 100 percent. Men can't work effectively in these conditions, so we spent $3 million to build an air conditioning system. It delivers 850,000 cubic feet of fresh, cooled air into the workings every minute.

So you can see that even if a big deposit of gold were discovered tomorrow, there wouldn't be any gold coming out of it for many years. This means the supply situation in gold will continue tightening—and upward pressures on the price of gold will grow stronger.

PCCE: What's your outlook for the price of gold?

HENSHAW: The principal factor affecting the price of gold will continue to be the rate of inflation—and I don't know any country in the world that plans to stop inflation. Within a few years, I expect gold to exceed $200 per ounce.

PCCE: Do you think now is a good time to buy gold?

HENSHAW: Yes, I believe those who purchase gold bullion, gold coins and gold shares will do well whether we go through continued inflation, a deep depression or some kind of international financial disaster.

11

America's Worst Depression Has Already Begun

Harry Browne

Harry Browne's first book, How You Can Profit from the Coming Devalution, *directly contradicted the conventional "wisdom" of Wall Street. Published in 1970, it warned that inflation would worsen; that wage and price controls would not work; that controls would lead to shortages; that the dollar would be devalued for the first time since 1934; and that gold and silver prices would soar.*

Now, four years later, those who followed Mr. Browne's advice—primarily to invest in gold, silver and certain foreign currencies—are all smiles. Profits from his sample investment programs range from 100 percent to 500 percent or more. Now he is no longer controversial; instead, his advice is sought by many who once thought his forecasts couldn't possibly be right.

A self-educated man, Harry Browne's original economic theories are compatible with the so-called "Austrian School" of economics—among whose adherents are Dr. Murray Rothbard, Dr. Franz Pick, and the late Prof. Ludwig von Mises.

Since 1967, Mr. Browne has been an investment consultant, helping individuals work out tailor-made investment programs to cope with the present crises. His current fee is $2,500 for a four-hour consultation.

He met Louis Carabini, president of PCCE, in 1964, and

the two have been close friends and associates since then. Mr. Browne was head of marketing for PCCE for a brief period during 1970; since then, he has made occasional lectures and TV commercials for PCCE.

In addition to his 1970 book, he is the author of two others. In How I Found Freedom in an Unfree World *(1973), he revealed how he learned to earn more, worry less, and enjoy life more. Richard Bach, author of the runaway bestseller* Jonathan Livingston Seagull, *called Browne's book "a gift of power and joy to whoever yearns to be free."*

Mr. Browne's third book, published in January 1974, is You Can Profit from a Monetary Crisis—*becoming the No. 2 bestseller nationally. The book expands on economic and financial concepts contained in his first book; reviews what has happened in the past four years; and updates his interpretation of the monetary crises that are affecting us all. If any book is "must" reading for serious investors, this is it. The present interview was published in March and April 1974.*

PCCE: In Nixon's State of the Union message in January, he promised "There will be no recession in the United States of America." Would you comment on that?

BROWNE: He's right. There will be no "recession." It is a depression—and it has already begun, will get worse and will last many, many years. I believe history will record January 24, 1973, as the date upon which it started. That day the Swiss government quit supporting the U.S. dollar; and within a few days all other major governments did likewise.

This allowed the dollar to fall to its natural marketplace level—which caused the prices of *all* products and services in the U.S. (imported or domestically-produced) to go way up in price. What the politicians overlook is that when a currency drops in value, domestically-produced goods can be sold for higher prices overseas—and so U.S. consumers must pay more for them in order to keep them from being exported.

As a result of the price increases of 1973, most Americans are no longer able to maintain their previous standards of living. That situation will get worse during 1974, and even worse during 1975. And as a result of price controls, shortages abound.

PCCE: Are you saying that prices will go up even further this year?

BROWNE: Yes, definitely. The Dow Jones Commodity Index rose only 55 percent from 1925 to 1972. Then it *doubled* in just *one year*—from August 1972 to August 1973. The index covers all major commodities—metals, grains, and livestock—at the producer level. Those increases are now being felt at the wholesale level. I expect to see a severe retail price explosion sometime between now and June.

PCCE: Some people might ask, "Can't the government stop this with price controls?" How would you answer this?

BROWNE: No—price controls can't cover everything, and as long as the government keeps printing paper money, the money will be spent on something. That causes the prices of *un*controlled products to skyrocket. And that, in turn, makes it unprofitable to produce products that *are* price-controlled—causing shortages.

PCCE: And would you explain to our readers what would happen if the government stops printing new paper money?

BROWNE: Then we will have a gigantic crash soon afterward—within six months or so. The government needs to continue inflating in order to pay its own bills as prices rise. To stop inflation means to cancel government contracts, cut welfare payments, end many subsidies. The government believes it can provide a little bit of inflation to boost the economy, but "a little inflation" is almost impossible.

You see, once the government starts inflating the money supply, it's hard for them to stop. A slow down would sign the death warrant for many companies that have been kept going by government subsidies. Bank loans would go into default, withdrawals to pay bills would abound—causing a massive liquidity crisis which would cause many people to have to sell their stocks, mutual funds, and real estate.

IS A CRASH INEVITABLE?

PCCE: Are you saying, then, that we will have a crash?

BROWNE: Not necessarily. When retail prices explode this spring, the government will be faced with the same problem that faces the rest of us: How to finance one's budget with prices much higher. The government will have two choices—(1) cut its budget; or (2) continue its spending programs and finance the higher costs with a much larger deficit (perhaps 20 to 30 billion dollars), paid for with printing press money.

If it chooses the first route, it will knock the props out from under many businesses and individuals that have been depending upon government aid for survival. This will cause a liquidity crisis—leading to a crash probably sometime around the end of 1974 or the beginning of 1975.

PCCE: And the second route?

BROWNE: If the government tries to maintain all its spending programs by resorting to the printing press, it would have to increase the paper money supply by ten percent or so within a six-month period—an unprecedented injection of inflation.

That, in turn, will cause prices to go much higher yet—causing another crisis point later in the year. That, in turn, will require another injection of inflationary paper money by the government to keep things going. And that causes another price explosion, etc., etc.

You see, for many years the government could inflate periodically on an apparently harmless basis. Inflation caused problems—but they weren't noticeable, especially to the government. Now, however, we've reached the point of no return. If the government stops, we'll have a crash. If they continue, the doses of inflation will get closer and closer together.

If the government doesn't back off, we could reach runaway inflation by the end of 1975. At that point, the government would have the printing presses going 24 hours a day and prices would be changing daily. It's no longer an abstract possibility; it's right around the corner.

PCCE: Which way do you think the government will choose?

BROWNE: I'm not betting either way. It's possible to make investments that will protect you whichever way it goes. And there's a good chance it will go both ways.

For example, the government could cut its budget this spring, recoil at the consequences that causes, and then turn on the printing press full blast. The inflationary input might be too late to prevent the crash but still cause a runaway inflation. Or it could continue to inflate but not inflate fast enough, still bringing on a crash.

In general, you have to assume that the politicians and bureaucrats will cut spending only as a very last resort. The Federal Reserve Board may try to hold the executive branch back by refusing to buy government bonds (the source of printing press money), but I would assume that the FRB would back off in a showdown.

So I don't expect budget cuts. But it's also possible that the bureaucrats may not realize how much inflation will be necessary to protect the fruits of past inflation. In that case, the crash will come regardless.

AN "INFLATIONARY DEPRESSION"

PCCE: You said we were already in a depression. Would you expand upon that point?

BROWNE: A depression is a period in which most people can no longer maintain their previous standards of living. That can be caused by higher prices, lower wages, investment crashes, unemployment, shortages, even wars. Whatever the cause, the result is the same—most people have to reduce their standards of living. But unless there's a crash, anyone can be convinced there's no depression—even if his standard of living is way down. He just assumes his problems are unique.

I doubt that more than 25 percent of the population is

living as well now as a year ago. The rest have not increased their dollar earning enough to compensate for the shortages and much higher prices.

Unfortunately, as I've already indicated, the situation will be much worse in 1974. While prices of important items probably increased about 20 percent during 1973, the price increases for 1974 could be at least that much by June. And then there'll probably be another price explosion later in the year.

If the 1973 problems could somehow be reversed during 1974, history would probably call 1973 only a difficult year or a minor recession. But there's no way to reverse those problems. One way or another—through much higher prices or a crash—1974 will be worse. And 1975 even worse. And someday historians will look back to see when the depression began. I believe the starting date will someday be assumed to have been January 24, 1973.

However, I'm sure that as matters get worse during 1974, the politicians will continue to talk about "preventing a recession" or "keeping the present prosperity going."

Inflation allows irrelevant indexes like the "Gross National Product" to continue going upward. So, in the eyes of most people, how could we be in a depression? What we have, however, is an *inflationary depression.*

PCCE: What will the inflationary depression do for most investments?

BROWNE: This unique phenomenon makes it harder and harder for the individual to know where he stands. I expect stock, mutual fund and real estate prices to go higher during 1974. But they won't be rising as fast as the cost of living. Anyone investing in those areas will probably lose purchasing power—just as he's been losing purchasing power for the past eight years.

Anyone who purchased the Dow Jones Industrial Averages in 1966 now has a paper loss of about 15 percent. But, in addition, the dollars involved have lost at least 30 percent (at least!) during that time. So, if he were to sell his investment now, the dollars he'd receive would purchase for him no more than 55 percent of the products and services he could have bought with the invested money in 1966.

That situation will accelerate during 1974 and 1975.

In terms of purchasing power, you'll have all the normal characteristics of a depression. Stocks, mutual funds, real estate, and the luxury investments like art, diamonds, numismatic coins, etc., will be going up in price—but not as fast as the prices of food, clothing, and other necessities. And so, even though the investments will be losing purchasing power, their owners may not realize the losses they're taking.

Commodities will probably take off upward again later this year (although I don't recommend them as investments for most people—as we'll discuss later). Probably the only investments that will move substantially faster than the increases in living costs will be gold, silver, and gold-backed currencies—as we'll also discuss later.

PCCE: How does the oil shortage affect all this?

BROWNE: It doesn't. Oil is just one more commodity that has to go up in price—to keep even with the ever-increasing supplies of paper money that are bidding for its purchase. The U.S. oil shortage is wholly the result of government intervention.

Supply and demand considerations constantly change for all commodities—resulting in continual price changes. If demand increases, or supply dwindles, the price goes up. But that doesn't mean that everyone will pay the higher price. Each person decides for himself how much he needs in the way of oil products at the current price. He cuts back, as needed, in ways of *his* choosing.

Rationing and allocations only happen when the government is determined to hold the price down—causing people to continue to try to buy at a level consistent with an unrealistically low price.

In most European countries, all rationing and driving restrictions have been abolished—and the price has been left free to move upward. Consequently, the price of gasoline has gone up about 10¢ per gallon, and there are no lines at gas stations, no shortages at all. Obviously, most people are using less oil at the higher prices—but each person has cut back in ways of his own individual choosing.

PCCE: We've often heard that the oil problem will hurt Europe more than the U.S. because America is largely self-sufficient in oil. Is that true?

BROWNE: No. Typically, the financial experts have overlooked many of the consequnces of an act.

Oil is a worldwide commodity. Prices in all parts of the world rise to a somewhat common level. Otherwise, all the oil is sold wherever the price is highest—and people do without elsewhere. In reality, people everywhere bid the price up to satisfy their purchasing needs. Prices will be just as high for a motorist in Saudi Arabia as in Holland, the U.S., or anywhere else.

If anything, people in the U.S. will be hurting more than anyone else if the government doesn't abolish the price controls on oil.

PCCE: You're saying, then, that the U.S. dollar will fare no better in the oil crisis than the European currencies.

BROWNE: Yes.

PCCE: Why, then, did the dollar appreciate against the European currencies last fall?

BROWNE: Because the Federal Reserve System borrowed foreign currencies from the Swiss, Belgian and German governments to buy dollars in the free market—pushing the dollar upward in price. But, as with all government programs, that only delayed the consequences. The borrowed currencies had to be returned to their owners. And the only way that could be done was for the Federal Reserve to go into the market in February to buy them, selling dollars. And so the European currencies started rising in price again.

U.S. BANKS AND SWISS BANKS

PCCE: What do you believe will happen to the dollar during 1974?

BROWNE: I believe it will continue to depreciate against the gold-backed currencies—the Swiss franc, Dutch guilder,

Belgian franc, Austrian schilling, and German mark. And I believe that investing in the first two of these currencies is a good way to compensate for purchasing-power losses in the dollar.

PCCE: Since most American banks are unable to handle deposits in those currencies, are you suggesting that this be done through a Swiss bank?

BROWNE: Yes. There are three main reasons for using a Swiss bank: (1) to get your money out of dollars and into a safer currency; (2) to keep your money outside of the country you live in, and so out of reach of the government; and (3) to deal with a safer banking system than what exists in the U.S.

PCCE: What's your view of banks in the United States?

BROWNE: I'm afraid I have no faith that banks are any safer now than they were during the 1930s. Inflation is much worse now than it was during the 1920s and 1930s. That means that banks are more vulnerable to liquidity problems in case of a crash.

Furthermore, I think people are being lulled into a false sense of security about the safety of their bank deposits through what I call the "sticker principle."

Just because a bank has a sticker in the window saying the deposits are insured by the government doesn't mean that the government has the assets to cover a string of bank runs. I doubt if the Federal Deposit Insurance Corporation could meet a run on even one large American bank.

The FDIC insurance fund has only $4.7 billion to cover insured bank deposits of $393.3 billion (as of the end of 1971.) And of the $4.7 billion, all but $10 million is invested in government bonds! The FDIC is just another way of financing government deficits. The Federal Savings & Loan Insurance Corporation is in similar shape.

So the government is relying on the sticker in the window to prevent bank runs. It won't, however, because bank runs aren't caused by panics; they're caused by liquidity problems. In the same way, crashes are caused by liquidity problems, not by panics.

PCCE: Would you explain that?

BROWNE: Yes. If you look back through history, you find no examples of crashes or bank runs that were caused by panics. The panic always *follows* the crash. The crash is caused when the government can no longer maintain its previous rate of inflation—causing bank loans to be called and tremendous demands for cash.

This causes great numbers of people to sell their stocks and make demands upon banks for loans and cash. Those people may feel very confident about the future of the economy; they assume that their problems are wholly individual. But when many people do the same thing, you have a crash or a bank run.

In the stock market, that causes a panic—because you can see the prices plummeting. But the demands for cash don't cause bank runs because most people are unaware of what's going on—until a large bank fails because of the liquidity crisis. Then the runs start—with or without deposit insurance. Bank holidays follow, which makes most people that much more anxious about their money.

Because the government cannot avoid runaway inflation without a massive liquidity crisis, I wouldn't care to keep any more money in an American bank than I could afford to lose.

PCCE: What would happen if Americans chose to follow your approach?

BROWNE: They won't. But even if they did, massive withdrawals couldn't hurt a soundly managed bank. And those that aren't soundly managed will go under eventually anyway.

ANY ALTERNATIVE TO DISASTER?

PCCE: You've spoken of the possibilities of a crash, runaway inflation and bank failures. Do you believe anything can be done to prevent all this?

BROWNE: Nothing. You can't inflate a currency for 40 years and then expect to avoid the consequences. It would require a rewriting of four decades of history—and I don't know anyone who could pull off that kind of miracle. Every act has a consequence, and you can't expect the government to inflate without consequences.

Now we have reached the point where the consequences are no longer a far-off, vague threat; they are already a present reality. America's worst depression has already begun and will be with us for a long time.

The only choices that the government can make now are those that will determine whether or not we have a runaway inflation—in addition to the depression that is already a reality.

PCCE: Given the premise that you can't foretell what the government will do, how likely do you think it is that we'll have a runaway inflation?

BROWNE: I see it as a 75 percent probability within the next three years—and probably by the end of 1975, or sooner.

PCCE: Does that mean the currency will become totally worthless?

BROWNE: No, not necessarily. But it does mean that assets in fixed dollar amounts—bank deposits, life insurance policies, Social Security, bonds, etc.—will be relatively worthless. Whether or not we will have a total currency collapse will depend upon what the government does during the runaway inflation.

Once you have prices changing daily in products that are normally far less volatile—hard goods like furniture and appliances, and in clothing, etc.—there will be no way for the government to halt it. The only thing the government could do would be to introduce a new currency that is redeemable in gold by anyone—otherwise, no one would have any reason to trust the new currency more than the old one.

If the government squanders its gold supply before then, in a futile attempt to suppress the free market gold price, it will be helpless to stop the runaway inflation. In that case, it will lead to the total destruction of the currency. I see that as a 50 percent possibility.

PCCE: What would happen then?

BROWNE: You'd have the worst kind of a depression.

During a normal depression, the economy is in terrible shape but the structure remains intact. If the currency collapses, the structure itself goes down. No one could employ anyone else without a medium of exchange. You couldn't make a phone call or mail a letter; you'd be immobilized.

If that happens, individuals would be reduced to a state of cautious barter—without protection or knowledge of events more than a mile away.

PCCE: And that's unavoidable if the currency collapses?

BROWNE: Not entirely. There's one possible way the structure can stay intact without a currency. If, *during* the runaway inflation, many people start using silver coins and gold coins in order to maintain constant price levels, trade can continue after the currency collapse—in gold and silver coins. You see, you can only use such a medium of exchange when you have some idea what a silver coin is worth relative to other things.

If the currency collapses without coins having been used as an alternative medium, no one will know how much he should give up in exchange for a silver dime—because he won't know what he can buy with it. But if prices of commodities have already been established in terms of coins, and if a widespread number of people and employers have coins, trade can continue somewhat uninterrupted.

Without that, it could be many years before some semblance of order is reestablished in the country. And that's another reason for having wealth outside the country—so that it will remain intact during such a difficult period. When order is reestablished, the wealth you've preserved might be enough to last you for life.

PCCE: Can you provide some broad investment principles to apply to the situation?

BROWNE: First, recognize that you can't know exactly how things will unfold; there are too many variables. So it's important to hedge in three directions: against a crash, against a runaway inflation, and against total currency collapse.

Remember that every crisis presents opportunities. There are many, many people who have made great sums of money during the past three years of crises—only because they saw them coming and acted accordingly. The same will apply to the next few years.

No one need feel helpless. There's still time to get your house in order, get out of vulnerable investments, and into the investments that will preserve and enhance your wealth —however large or small it may be.

The world isn't coming to an end, even though a lot of what we've taken for granted will be coming to an end. But there will still be much to enjoy and many good reasons for being alive. So don't take a defeatist attitude; exert the energy necessary to protect yourself.

And then you can sit back and let things unfold as they will, without having to feel that everything you read in the newspaper is a threat to your existence.

FOUR POSSIBILITIES TO PROTECT YOURSELF AGAINST

PCCE: Would you summarize for our readers what an investor should protect himself against?

BROWNE: An individual should be protected against four basic possibilities. Each of them is a very real possibility, and each could cause great harm to an individual's investments if he hasn't hedged against them.

The first is *deflation*. If the U.S. government *doesn't* try to keep up with the higher prices by escalating the printing of paper money, there will be a liquidity crisis. It may not even be an actual deflation—in the sense that the paper money supply would be diminishing. But if the government's inflation doesn't keep pace with itself, there will still be liquidity problems—bank failures, investment crashes, etc.

Second is *runaway inflation*. If the government *does* escalate the inflation—in order to maintain its spending programs at the higher prices caused by its own inflation—it could lead to runaway inflation by the end of 1975. Then paper money could be relatively worthless.

Third is *civil disorder*. If there's a runaway inflation, it could lead to a total currency collapse—accompanied by an end to a medium of exchange, no normal employment, no products or services available. Once a runaway inflation starts, this will be avoided only if the government introduces a new currency convertible by *everyone* into gold, or if individuals have already started exchanging silver coins and gold coins and have thereby established workable price levels in gold/silver media.

And the fourth is *government intervention*. No matter which of the first three possibilities occurs, we can expect the government to intervene aggressively. Any investment in the U.S. will be subject to new government regulations, excess profits taxes, price controls, and other possibilities we may not even be able to imagine now.

Each of these possibilites is so real that any prudent investor should insulate himself against all of them. Many gold-oriented investment counselors have been arguing over deflation and runaway inflation—trying to demonstrate that we will have one or the other. Such arguments are pointless. No one can foretell which way things will go—and we may even have both. And since is is possible to protect against all these possibilities, it's foolish to try to outguess the future by hedging against only some of them.

PCCE: You've made it clear in *You Can Profit from a Monetary Crisis* that you don't think the traditional investments can hedge against all four possibilities.

BROWNE: No, they can't. "Growth" investments such as the stock market, mutual funds, real estate, collectors' items, etc., would be hurt badly if there's a deflationary crash. Also, I doubt that they'd appreciate fast enough in a runaway inflation to keep up with the higher cost of living. So you'd be losing purchasing power even while you're showing paper profits.

Meanwhile, fixed dollar investments such as bank accounts, certificates of deposit in U.S. banks, bonds, mortgages, etc., would be virtually wiped out in a runaway inflation.

And all these investments are highly vulnerable to government intervention and civil disorder.

For several years, I have *not* recommended any of these investments for any of my clients—unless an individual had enough money for him to protect himself, plus had some money he wanted to keep in the U.S. economy.

PCCE: Which investments do you believe will protect against all four possibilities and provide profits?

BROWNE: Basically, gold, silver, and gold-backed currencies.

PCCE: Are these the only investments you believe have a chance?

BROWNE: No. There are bound to be other things that will appreciate faster than the dollar depreciates and will, in retrospect, prove to have avoided the dangers. But they're too chancy to bet on in advance. And if one invests in gold, silver, and gold-backed currencies, he's going to have all the protection and profit he needs.

PCCE: How about platinum or other precious metals?

BROWNE: I'm often asked about platinum, plutonium, other precious metals, and other commodities that are currently in short supply. But the three recommended investments have advantages that these other investments don't have.

For example, the shortages of the *recommended* investments have been caused by governments—mainly the U.S. government—over long periods of time; they aren't temporary shortages that may change soon.

Also, no specialized knowledge is necessary—beyond the minimal amount you would need for whatever you intend to do with your money.

What's more, these investments are all depression-proof, which most of the others are only inflation-proof.

Finally, you can arrange them so that you don't have to be the least bit concerned over any future U.S. government actions.

In one sense, there's no real shortage of anything in the world—while there's a tremendous oversupply of U.S. dollars. It only seems to be a shortage when the price is artificially held down so that the demand at that lower price is greater than the supply. When prices are free to move upward, less urgent users of the commodity make do with something else and eventually forget about their demand for the original commodity.

So-called shortages only appear to be a national issue when a government has forced the suppliers to charge an artificially low price and thereby inspired an unrealistic demand.

For example, there's no shortage of Cadillacs in the world—even though very few people have them. Very few people try to buy a Cadillac at its present price. But if Cadillacs sold for $2,000 apiece, there would seem to be a gigantic shortage.

Recent price controls have created apparent shortages in many things—but those shortages can be reversed quickly by an end to price controls. But not so with gold, silver, and gold-backed currencies. Because in each of those cases, the shortages are the results of years and years of price controls.

So I only advise clients to get into other things when one of two situations exists—the client is willing to devote the time to learning about that investment and supervising it daily or he has extra money to play with that he's quite willing to lose if things don't work out for him.

PCCE: Let's talk a little more about your three recommended investments one at a time—starting with gold.

GOLD

BROWNE: Gold was price-controlled for many years because the U.S. government—and other governments—wanted to maintain the fiction that $35 equalled one ounce of gold—in other words, that "the dollar was as good as gold."

Anyone who had gold to sell was free to ask any price for it he chose. But as long as governments guaranteed to sell it at $35 per ounce, no one had to pay more for it—and the price was effectively held at $35.

It's important to realize that if the U.S. government hadn't inflated its currency, it *could* have and *should* have continued paying out gold at $35 per ounce indefinitely. But instead, it *did* inflate its currency by issuing more than 35 dollars for each ounce of gold it had. So it couldn't possibly continue paying out gold at $35 per ounce.

As of January 30, 1974, there were 1,250 dollars in circulation (that's checking account deposits plus cash outside of banks) for every ounce of gold in the U.S. Treasury. To be honest, then, the government should withdraw dollars from circulation until the ratio is $42.22 with gold, or it should acknowledge that each ounce of gold is equal to 1,250 dollars. Naturally, it does neither.

In modern practice, a government can get away with having only 25 percent or 30 percent of the gold necessary to back all its currency. But the U.S. government has only about three percent—not nearly enough to withstand a run on its gold. As a result, it refuses to pay out any gold at all.

We can see how disastrous the government's policy of the 1950s and 1960s was. First, it inflated its currency energetically—so that the ration of dollars to gold climbed higher and higher. Second, it squandered what gold it did have by selling it in the free market at an artificially low price—further widening the ratio. For close to ten years it tried to hold the price down.

Eventually, of course, the government became powerless to hold the price down any longer. It's important to realize that no government is omnipotent—because no government can have the resources to stop all individuals from pursuing their self-interests.

And so when the gold price was free to move, it had many years of lost time to make up. So far, it has quintupled—but with over a thousand dollars in circulation for every ounce of gold backing them, the gold price has a long way yet to go.

PCCE: How would gold be affected by the four possibilities you outlined earlier?

BROWNE: Let's run through them.

As for *deflation,* a deflation-type depression isn't likely to terminate gold's longterm price rise; but it could cause a temporary selloff by unwise holders of gold who are concerned only about a runaway inflation. Soon after, however, gold would go higher again.

The government would have to deflate the currency by at least 30 percent to make the gold price level off at $200. Obviously, that won't happen. If a deflation-type depression occurs, it will be only because the government hasn't inflated fast enough; there probably won't be *any* actual deflation. So in the midst of the crashes of other investments, gold will still have to continue upward to at least $300, if not more.

In a *runaway inflation* gold will have to continue upward in price to keep up with the disparity between dollars issued and the gold supply. However, gold has additional ground to make up because of the years of price controls. So you can count on it to appreciate more than the cost of living.

In addition, it will become progressively more difficult to settle contracts in dollars during a runaway inflation. So, more and more, individuals will be turning to gold as the only unvarying means of pricing longterm contracts.

If we have *civil disorder,* gold coins may be useful to make large purchases after—and if—the currency collapses. Otherwise, gold will have no immediate usefulness *during* such a period. But it will be the best way of keeping wealth intact while awaiting a return to normality; for you can be sure that gold will always be at the root of any future monetary system.

GOLD STOCKS AND GOLD COINS?

As for *government intervention,* here the important issue is *how* you invest in gold. In my 1970 book, *How You Can*

Profit from the Coming Devaluation, I recommended gold stocks—because gold bullion was illegal for Americans and the gold coin market was too thin to count on. Now, it's still illegal to own bullion—but the coin market is big and useful.

As a result, I no longer advise gold stocks because they are too vulnerable to governmental intervention. In the U.S., there's an all-out war on between the government and anyone who makes sudden profits—or almost any kind of profits. In Canada, almost any mining venture is either subsidized or penalized; "free enterprise" is unknown except for campaign oratory. And the South African government is even more restrictive economically than the U.S. government; there's no reason to assume that government will let the mining companies operate unhindered forever.

So if it's safety and legality that you want, gold coins are the only answer I can find. If you buy them over the counter in this country, you can easily store them in a way that's invulnerable to any future governmental decree. Or you can buy them through a Swiss bank that will store them for you in complete privacy.

PCCE: The two normal arguments in favor of gold stocks are the built-in leverage many have and the income provided by dividends. Are these advantages worth facing the risks you cited?

BROWNE: Not in my opinion, but I have no argument with anyone who knows the risks and still wants to take them.

The leverage isn't that unique. Anyone who wants leverage can buy gold coins on two-to-one leverage—50 percent margin—and outgain most any gold stock.

To do better, you have to look for a "penny" gold stock that has low earnings at a $175 gold price. The mining company might then multiply its earning many times over when gold reaches $250. So you might triple or quadruple your investment with only a 40 percent increase in the gold price.

Taking a gamble like that makes more sense to me than

investing in an already-established gold mine. The latter doesn't offer enough to offset the risks. But any investment in gold mines should be treated as a pure gamble—to be taken only with money you can afford to lose—because you're too vulnerable to a greedy government. Gold mines can't operate in privacy.

PCCE: What about gold stocks for income purposes?

BROWNE: This is a very important point. I believe that the search for income should be abandoned for at least the next two or three years—until one can see just how the scenario is unfolding.

My advice to anyone—including retired people—is to ignore dividends and interest when picking investments right now. Choose all investments by your own standards of safety and profit—not by income. In general, you can only get income now by jeopardizing your capital.

Far better to *draw* on the capital for awhile, while making sure that it's safe. I realize this goes against the age-old tradition, "Live off the income, not the capital." But the times right now call for a different approach.

Forego the five percent per year income and deplete your capital by 15 percent over three years instead. Otherwise, you risk losing 50 percent or more of the capital forever. And the most likely possibility is that you'll be drawing on capital *appreciation,* so that you'll still have at least 100 percent of what you started with three years from now.

PCCE: Which of the most popular gold coins do you recommend?

BROWNE: The Mexican 50 peso. It has a large, tested market, and it has a much lower premium over the bullion price, compared to other popular coins.

SILVER AS AN INVESTMENT

PCCE: What about silver? Are you counting on it as a monetary metal?

BROWNE: Yes and no. There are really two separate

investments involved—silver bullion and silver coins. The latter has monetary potential, but the former doesn't.

In discussing monetary systems, one should always keep in mind that the marketplace establishes a monetary commodity gradually; no government can successfully impose a commodity as money upon an economy where individuals themselves haven't found it to be the most useful.

For thousands of years, gold has been found to be the most useful monetary commodity by far. Silver has been in second place—but a distant second—useful only in special circumstances. And I should add that whatever is in third place is way behind silver.

So I'm not looking for the price of silver bullion to go up because of the monetary crises. Rather, silver will continue to become more expensive because of the artificial shortage created over a 100-year period by the U.S. government. The details of that shortage are summarized in my recent book, *You Can Profit from a Monetary Crisis*, so I'd rather not use the space here to repeat them.

The important point is that silver must go up in price, more than most commodities, to offset many years of price control—in addition to the more normal current factors of heavy inflation and recent government price controls. So silver is an ideal way of keeping way ahead of higher living costs.

On the other hand, if there's a total currency collapse, silver *coins* will probably be the first commodity to replace barter and become money. This is because the coins are already minted, are easily recognizable for what they are, and are in smaller value denominations than gold coins. They are a ready medium of exchange to fill the void.

Of course, without a currency collapse, the price of silver coins will go up approximately as much as silver bullion goes up. Thus, if you buy them as a hedge against a currency collapse, you'll also profit if the worst doesn't happen.

PCCE: How effectively do silver bullion and silver coins protect assets against the other three possibilities you outlined?

BROWNE: Mostly, both should react in the same way.

First, let's consider *deflation.* If there's an orthodox depression, signalled by a crash of most investments, I would expect silver to be sold in panic by the fainthearted. However, the price should recover and eventually go even higher than it would have without the depression. Industrial consumption of silver would undoubtedly be reduced; but the *production* of silver would be reduced even more—since it relies so heavily on the production of copper, lead, and zinc, traditional depression losers.

In addition, silver is already underpriced—in terms of actual consumption and production—because of the many years of price control. The annual production deficit is still being offset partially by sales of silver from those who bought it as a bargain under $2 per ounce. Since there isn't enough silver available from that source, consumption is already dropping—as former silver users are being weeded out by price increases. But the silver price will have to go much higher to force annual consumption down to the level of annual production.

As for *civil disorder*—if the bullion is kept in Switzerland, it can provide a store of wealth to wait out abnormal times. The coins are intended as a potential purchasing medium so they must be kept closer to you in a safe place.

Then there's *government intervention.* Bullion is out of the reach of any problems if kept in a Swiss bank. No Swiss laws have been passed in the past two generations that have interfered with any investments already held by foreigners or citizens; the situation is the opposite of the U.S. where any new executive or new whim can overturn the past.

If silver coins are purchased over-the-counter here and stored safely, they are out of the reach of the U.S. government. Coins are vulnerable to government intervention only if stored with a coin dealer or in a silver warehouse.

Again, one of the advantages of silver is that you can invest in it without making yourself vulnerable to future government whims.

CURRENCIES AS AN INVESTMENT

PCCE: That leaves only gold-backed currencies. Why are they useful—especially if gold and silver provide adequate protection?

BROWNE: I believe that the holding of some currency is necessary to provide a well-rounded investment program.

First, I never find it profitable to overestimate my own knowledge or foresight. Diversification is almost always a good thing—so long as you diversify within the limits of the general principles you're following.

Second, currencies provide short-term liquidity. If you hold gold or silver for three years or more, the chances are probably less than 10 percent that you could suffer a loss. But in the *short-term,* your risk is much, much higher. Both gold and silver are five times higher than they were a few years ago. That leaves a lot of room for short-term price fluctuations. If you need some dollars six months from now for an emergency, you don't want to have to sell these commodities if they both happen to be 25 percent or so below the prices at which you bought them.

A good currency can provide the necessary short-term liquidity. For instance, at 33¢, the Swiss franc has a realistic downside risk of about 4¢—or 12 percent. You won't have stomach pains if you should need to cash in some of them at a small loss to raise dollars for current needs.

This is especially useful for the individual who needs to draw on capital for awhile to replace the income he received from investments now discarded. Also, the individual whose occupational income isn't keeping up with inflation can augment his income by drawing on Swiss franc capital until he has sizeable profits in gold and silver.

PCCE: Do you see much profit potential in foreign currencies?

BROWNE: Possibly, yes. Currencies are only substitutes for gold. As currencies fail to perform that function properly, individuals turn more and more to other currencies that *are* proper substitutes or to gold itself.

To be useful, a currency must be redeemable for gold by anyone who wants it, and it must be redeemable at a rate that's permanent and can be depended upon by those who received the currency. Today, I know of no currency that is redeemable for everyone—but several currencies have the gold backing to make that possible.

The current international monetary chaos will subside only when the major currencies can be exchanged for gold—by governments at least. It's doubtful that this can be done without a minimum of 20 percent gold backing at the official rate. The U.S. dollar has only three percent backing at the official rate of $42.22 per gold ounce. So a minimum useful devaluation would have to be 85 percent—which would change the gold redemption rate to $250 per ounce and provide 20 percent gold backing.

At the same time, however, there are currencies that already have backings of 20 percent, 25 percent, or more. The Swiss franc, for example, had a gold backing of 24 percent at the end of 1972, which is the latest year I have figures for. The Swiss government could *up*value its currency—making it *more* valuable—and still have 20 percent backing.

To make the currencies redeemable in gold, various devaluations—and possibly upvaluations—will have to take place. This, in turn, will create new exchange rates between the currencies. Those that were stronger originally will command much higher prices in terms of the weaker currencies. In other words, a Swiss franc will be priced much higher in dollars.

PCCE: What if the governments *don't* follow that course of action?

BROWNE: That will mean that the weaker currencies won't become redeemable—and so they will become progressively less and less useful as money. Again, the stronger currencies will have to cost more; otherwise, no one would trade a strong one for a weak one.

PCCE: Do you feel there is a likelihood that governments—like the Swiss, perhaps—will be tempted to depreciate their currency in order to protect export trade?

BROWNE: That's certainly a possibility. Avoiding that temptation requires sound monetary management, and there's not much of that around among governments today.

A government can only keep its currency inexpensive by inflating it. That keeps its exports competitive, but only by injuring the purchasing power of its own citizens. In other words, exporters are actually being subsidized by everyone else in the country. As always, free lunches have to be paid for by someone.

If a currency is sound, there's no reason to worry about its exchange rate with other currencies. As the sound currency appreciates, less *units* of products will be exported—but each unit of currency will then buy more units of imports. The only purpose of production or sales is to acquire things to consume—and the nation with a sound currency will be able to do this no matter how much its currency appreciates.

However, that doesn't mean the government of any given nation will see it that way. I have no faith in the permanent rationality of any government. So no currency has the assured long-term safety that gold has. But over the next few years, the Swiss franc and the Dutch guilder—because of their heavy gold backing—should appreciate in terms of dollars—even if the governments involved inflate them—because they have long headstarts over the dollar.

I should mention that Swiss banks are much less vulnerable to government mistakes than the Swiss currency is. Inflating the Swiss franc might create popularity for the government—but tampering with the banks would cause an uproar among probably 90 percent of the Swiss citizens. It will take another generation or two before the general Swiss fetish for privacy could be educated out of the culture.

PCCE: In your new book, you mentioned the Lebanese pound favorably. Would you recommend it, or any other currency, in addition to the Swiss franc and the Dutch guilder?

BROWNE: Generally, no. The Lebanese pound had 47 percent gold backing at the end of 1972—far and away the strongest major currency in the world. But because it's a very young currency and because of the political instabilities of the Middle East, it is a risky proposition. I recommend the Lebanese pound only for gambling money.

Other strong currencies, in terms of gold backing, are the South African rand, the Portugese escudo, the Austrian schilling, and the Belgian franc. Each of them has some political drawbacks, however, which we don't need to go into here. I think it's sufficient diversification to use two currencies—and, generally, I recommend that about 65 percent to 75 percent of one's *currency* budget be in Swiss francs and the rest in Dutch guilders. If a third currency is necessary for any reason, I would recommend the Belgian franc—if safety is the object.

Such currencies are best kept in a Swiss bank account. You can keep an account in a currency other than the Swiss franc. However, most Swiss banks will want you to hold up a Eurocurrency account if you want to own any currency other than Swiss francs; and Eurocurrency accounts are not as safe as regular bank accounts. The Foreign Commerce Bank is the only one I've found so far that offers a current or deposit account in Dutch guilders.

PCCE: How do these currencies stand up against the four possibilities?

BROWNE: Very well. Again, we'll look at *deflation* first. In a deflationary depression, currencies will increase in purchasing power. In general, the Swiss franc and the Dutch guilder should merely continue their appreciation against the U.S. dollar.

In a *runaway inflation*—as with gold and silver stored in Switzerland—you wouldn't want to exchange Swiss francs for dollars. Instead, they would preserve your wealth for a later date.

The only vulnerability here is the possibility that the Swiss government could get caught up in the hurricane of runaway inflation and cause the Swiss franc to suffer as the dollar would suffer. So, long-term wealth should be stored in gold and silver. Beyond anticipated cash needs of the next two years, I wouldn't recommend that more than 25 percent of one's assets be left in *any* currency.

As for *civil disorder,* as long as the currency is in Switzerland, it will be unaffected by any chaos here.

Then there's *government intervention.* The only vulnerability is to the possibility that the Swiss government might debase the franc—same for the guilder. Currencies in a Swiss bank are out of the reach of the U.S. government.

I believe that the Swiss franc and the Dutch guilder can play an important part in most any investment program—so long as you don't forget about them for several years. And the Lebanese pound can be an interesting speculation.

As with gold and silver, strong currencies have a lot of room to grow—and, again, because of years of price controls. From 1944 to 1973, the fixed exchange rate system prevented the Swiss franc and the Dutch guilder from appreciating to realistic price levels.

FINDING AND PICKING A SWISS BANK

PCCE: A Swiss bank is an important part of your recommendations. What guidelines would you suggest for finding and picking a Swiss bank?

BROWNE: If you travel to Switzerland, most any bank you visit can be utilized. But if you deal solely by mail, as most people do, you'll find that most banks will discourage your opening an account with them.

There are a few banks that *are* equipped to deal with nonresidents of Switzerland. All of the big three banks can handle foreign accounts. Of them, I like the financial position of the Union Bank of Switzerland much more than the others, and I stated that in the monetary book. As a result, the bank has been besieged with literally thousands of inquiries. And my contact at the bank has indicated that the bank officials are not particularly pleased by that. Therefore, I doubt that people writing to the bank now are getting much help.

The Foreign Commerce Bank has received a similar

volume, and its officers are much more willing to respond—although service has been slowed somewhat by the volume.

Here are a few additional guidelines—

In the first place, don't ever open a "discretionary" account—in which the bank will make the investment decisions for you. If you do, your money will probably wind up in American common stocks or bonds. Ask for advice, if you choose, but make your own decisions.

In addition, spell out your instructions carefully when you purchase anything. Don't be afraid to state the instructions a second time in different words—to be sure they're understood.

Lastly, if you make a purchase, ask the bank to cable you—at your expense—if, for any reason, the purchase won't be made immediately. I've seen many instances where a purchase couldn't be made for weeks afterwards. Many other guidlines are included in my new book.

AN INVESTMENT PROGRAM FOR YOU

PCCE: In your book, you devote two large chapters to the creation of an investment program appropriate to each individual investor. Can you summarize that advice here?

BROWNE: The investment program can be divided into two main categories. First of all, keep enough purchasing power with you to see you through a few years of no currency and civil disorder—in case that happens. This means silver coins and, possibly, gold coins. You should make arrangements to store them yourself.

Most families could probably survive for several years on three bags of silver coins, if necessary. If you don't have a safe way of storing more than that, all additional wealth to be kept here should be in the form of gold coins. How much you should have depends upon many individual variables.

Secondly, the rest of your wealth should be stored in a Swiss bank in gold coins (bullion, if you're unaffected by legal restrictions), silver bullion and the recommended currencies.

Suitable proportions of these three investments depend upon individual situations. You should keep in currencies those assets that you might need to convert to dollars in the next year or so. In general, gold is the safest long-term investment. Silver is the most volatile; it could show the largest dollar profit but is riskier than gold.

A sample program could have 50 percent in gold, 25 percent in silver, and 25 percent in recommended currencies. And you shouldn't let current gold and silver prices alter your overall investment program. Prices are not too high now to be buying for the longterm. But anyone who is not willing to hold it for a year or two could easily suffer a loss by buying now. At the same time there is no way to know if the price of either gold or silver will drop in the near term—and, if so, how far it would drop. So the best advice with long-term plans is to buy now and just wait things out.

But a cardinal rule is—*Never invest your assets in a way that leaves you uncomfortable.* Only you can know what worries you, what you hope to do, and how equipped you are to do it—taking into consideration your assets, your emotional liabilities, your family situation, and other factors. Keep juggling your program on paper until you find a format you feel comfortable with. Don't let anyone else make that decision for you.

PCCE: Thank you very much, Harry. Just one last question—we're pleased to note that *You Can Profit from a Monetary Crisis* was solidly entrenched on all national bestseller lists. Since your opinions and recommendations are so contrary to the prevailing views found in most news media, how do you account for the popularity of the book?

BROWNE: It appears that the news media, the politicians, and the economists are not in touch with much of popular opinion. In fact, the more the "thought leaders" assert that everything's going to be all right, the more concerned many people become. I'd even guess that the majority of Americans anticipate bad times ahead. However, only a very small percentage will take steps to protect themselves.

Fortunately, there *are* ways for an individual to protect

himself. If he's willing to take the responsibility for his own life, his own decisions, his own future, he has an excellent chance of coming out ahead when the washout is over.

Bibliography

BOOKS

Economics

Alchian, Armen A. and Allen, William R., *University Economics* (Belmont, California: Wadsworth Publishing, 1967)

Bastiat, Frederic, *Economic Sophisms* (Princeton, N.J.: D. Van Nostrand, 1964)

Baudin, Louis, *A Socialist Empire* (Princeton, N.J.: D. Van Nostrand, 1961)

Bresciani—Turroni, Constantino, *The Economics of Inflation* (New York: Augustus M. Kelley, 1968)

Browne, Harry, *How You Can Profit from the Coming Devaluation* (New Rochelle, N.Y.: Arlington House, 1970)

Browne, Harry, *You Can Profit from a Monetary Crisis* (New York: Macmillan, 1974)

Chamberlain, John, *The Roots of Capitalism* (Princeton, N.J.: D. Van Nostrand, 1965)

Chesnut, Mary Boykin, *A Diary from Dixie* (Boston: Houghton Mifflin, 1949)

Chou, Sun-Hsin, *The Chinese Inflation, 1937-1949* (New York: Columbia University Press, 1963)

Cliches of Socialism (Irvington-on-Hudson, N.Y.: Foundation for Economic Education, 1970)

Del Mar, Alexander, *The History of Money in America* (Hawthorne, California: Omni Publications, 1966)

Del Mar, Alexander, *Money and Civilization* (New York: Burt Franklin, 1969)

Diogenes, *The April Game: Secrets of an Internal Revenue Agent* (Chicago: Playboy, 1973)

Dulles, Eleanor Lansing, *The Dollar, the Franc and Inflation* (New York: Macmillan, 1933)

Dulles, Eleanor Lansing, *The French Franc, 1914-1928* (New York: Macmillan, 1929)

Einzig, Paul, *The History of Foreign Exchange* (London: Macmillan, 1970)

Ellis, Charles D., *The Second Crash* (New York: Simon & Schuster, 1973)

Erdman, Paul E., *The Billion Dollar Sure Thing* (New York: Charles Scribners Sons, 1973)

Fehrenbach, T.R., *The Swiss Banks* (New York: McGraw-Hill, 1972)

Flynn, John T., *As We Go Marching* (Garden City, N.Y.: Doubleday, Doran & Co., 1944)

Friedman, David, *The Machinery of Freedom* (New York: Harper Colophon Books, 1973)

Friedman, Milton, *An Economist's Protest* (New Jersey: Thomas Horton & Company, 1972)

Friedman, Milton & Schwartz, Anna Jacobson, *A Monetary History of the United States* (Princeton, N.J.: Princeton University Press, 1963)

Friedman, Milton, Ed., *Studies in the Quantity Theory of Money* (Chicago: University of Chicago Press, 1956)

Garrett, Garet, *The People's Pottage* (Caldwell, Idaho: Caxton Printers, 1965)

Grant, Richard W., *The Incredible Bread Machine* (Bray)

Greaves, Percy L., *Understanding the Dollar Crisis* (Belmont, Massachusetts: Western Islands, 1973)

Groseclose, Elgin, *The Decay of Money: A Survey of Western Currencies, 1912-1962* (Washington, D.C.: Institute for Monetary Research, 1962)

Groseclose, Elgin, *Fifty Years of Managed Money: The Story of the Federal Reserve* (New York: Spartan Books, 1966)

Haberler, Gottfried, *Inflation: Its Causes and Cures* (Washington, D.C.: American Enterprise Institute, 1966)

Harper, F.A., *Why Wages Rise* (Irvington-on-Hudson, N.Y.: Foundation for Economic Education, 1966)

Hayek, Friedrich A., Ed., *Capitalism and the Historians* (Chicago: University of Chicago Press, 1954)

Hayek, Friedrich A., Ed., *Collectivist Economic Planning* (London: Routledge & Kegan Paul, 1963)

Hayek, Friedrich A., *The Road to Serfdom* (Chicago: University of Chicago Press, 1944)

Hazlitt, Henry, *The Conquest of Poverty* (New Rochelle, N.Y.: Arlington House, 1973)

Hazlitt, Henry, *Economics in One Lesson* (New York: Harper, 1946)

Hazlitt, Henry, *Man vs. the Welfare State* (New Rochelle, N.Y.: Arlington House, 1969)

Hazlitt, Henry, *Time Will Run Back* (New Rochelle, N.Y.: Arlington House, 1966)

Hazlitt, Henry, *What You Should Know About Inflation* (New York: Funk & Wagnalls, 1965)

Hepburn, A. Barton, *A History of Currency in the United States* (New York: Macmillan, 1915)

Ilke, Max, *The Banking System of Switzerland* (Stroudsburg, Pa.: Dowden, Hutchinson & Ross)

Kellems, Vivian, *Toil, Taxes & Trouble* (New York: E.P. Dutton, 1952)

Kemmerer, Edwin Walter, *Inflation and Revolution: Mexico's Experience of 1912-1917* (Princeton, N.J.: Princeton University Press, 1940)

Kia-Ngau, Chang, *The Inflationary Spiral: The Experience in China, 1939-1950* (Cambridge, Massachusetts: Massachusetts Institute of Technology, 1958)

Kolko, Gabriel, *Railroads and Regulation, 1877-1916* (Princeton, N.J.: Princeton University Press, 1965)

Kolko, Gabriel, *The Triumph of Conservatism* (New York: Free Press of Glencoe, 1963)

Krefetz, Gerald, *The Dying Dollar* (Chicago: Playboy, 1972)

Lacy, Mary G., *Food Control During Forty-Six Centuries: A Contribution to the History of Price-Fixing* (Irvington-on-Hudson, N.Y.: Foundation for Economic Education)

Lane, Rose Wilder, *The Discovery of Freedom* (New York: Arno Press, 1972)

Lieberman, Jethro K., *How the Government Breaks the Law* (New York: Stein & Day, 1972)

Mises, Ludwig, *Bureaucracy* (New Rochelle, N.Y.: Arlington House, 1969)

Mises, Ludwig, *Human Action* (Chicago: Henry Regnery Company, 1966)

Mises, Ludwig, *Omnipotent Government* (New Rochelle, N.Y.: Arlington House, 1969)

Mises, Ludwig, *Planning for Freedom* (South Holland, Illinois: Libertarian Press, 1965)

Mises, Ludwig, *Socialism* (New Haven, Connecticut: Yale University Press, 1951)

Mises, Ludwig, *The Theory of Money and Credit* (New Haven, Connecticut: Yale University Press, 1953)

Morton, Frederic, *The Rothschilds* (New York: Atheneum, 1962)

Paterson, Isabel, *The God of the Machine* (Caldwell, Idaho: Caxton Printers, 1968)

Pazos, Felipe, *Chronic Inflation in Latin America* (New York: Praeger, 1972)

Peters, Harvey, *America's Coming Bankruptcy* (New Rochelle, N.Y.: Arlington House, 1973)

Pick, Franz and Sedillot, Rene, *All the Monies of the World* (New York: Pick Publishing, 1971)

Porteous, John, *Coins in History* (New York: G.P. Putnams Sons, 1969)

Rickenbacker, William F., *Death of the Dollar* (New Rochelle, N.Y.: Arlington House, 1968)

Riegel, E.C., *A New Approach to Freedom* (San Pedro, California: Heather Foundation)

Riegel, E.C., *Private Enterprise Money* (New York: Harbinger House, 1944)

Rothbard, Murray N., *America's Great Depression* (Princeton, N.J.: D. Van Nostrand, 1963)

Rothbard, Murray N., *For a New Liberty* (New York: Macmillan, 1973)

Rothbard, Murray N., *Man, Economy and State* (Princeton, N.J.: D. Van Nostrand, 1962)

Rothbard, Murray N., *Power & Market* (Menlo Park, California: Institute for Humane Studies, 1970)

Rothbard, Murray N., *What Has Government Done to Our Money?* (Colorado Springs, Colorado: Pine Tree Press, 1963)

Rueff, Jacques, *The Age of Inflation* (Chicago: Henry Regnery Company, 1964)

Rueff, Jacques, *Balance of Payments* (New York: Macmillan, 1967)

Rueff, Jacques, *The Monetary Sin of the West* (New York: Macmillan, 1972)

Schultz, Harry D., *Panics & Crashes and How You Can Make Money Out of Them* (New Rochelle, N.Y.: Arlington House, 1972)

Sennholz, Hans, Ed., *On Freedom and Free Enterprise* (Princeton, N.J.: D. Van Nostrand, 1956)

Smith, Adam, *The Wealth of Nations* (New York: Modern Library, 1937)

Snyder, Carl, *Capitalism the Creator* (New York: Macmillan, 1940)

Sobel, Robert, *Panic on Wall Street* (Toronto: Collier-Macmillan Canada, 1970)

Spencer, Herbert, *The Man Versus the State* (New York: D. Appleton & Co., 1897)

Stone, Willis E., *Where the Money Went* (Los Angeles: Fact Sheet, 1971)

Stucki, Lorenz, *The Secret Empire: The Success Story of Switzerland* (New York: Herder & Herder, 1971)

Sumner, William Graham, *The Financier and the Finances of the American Revolution* (New York: Augustus M. Kelley, 1968)

Sumner, William Graham, *What Social Classes Owe to Each Other* (Caldwell, Idaho: Caxton Printers, 1963)

Thomas, Dana L., *The Money Crowd* (New York: G.P. Putnam's Sons, 1972)

U.S. News & World Report, Inflation Simplified (Toronto: Collier-Macmillan Canada, 1969)

Weaver, Henry Grady, *The Mainspring of Human Progress* (Irvington-on-Hudson, N.Y.: Foundation for Economic Education, 1953)

White, Andrew Dickson, *Fiat Money Inflation in France* (Irvington-on-Hudson, N.Y.: Foundation for Economic Education, 1959)

Williams, Charles E., *Runaway Inflation: The Onset* (Brooklyn, N.Y.: Theodore Gaus' Sons, 1972)

Wooldridge, William, *Uncle Sam, the Monopoly Man* (New Rochelle, N.Y.: Arlington House, 1970)

Yeager, Leland B. and Tuerck, David G., *Trade Policy and the Price System* (Scranton, Pa.: International Textbook Company, 1966)

Gold

Allen, Gina, *Gold!* (New York: Thomas Y. Crowell, 1964)

Anderson, R.S., *Australian Gold Fields* (Sydney: D.S. Ford, 1956)

Baruch, Bernard M., *The Public Years* (New York: Holt, Rinehart & Winston, 1960)

Berton, Pierre, *The Golden Trail* (Toronto: Macmillan, 1954)

Blakemore, Kenneth, *The Book of Gold* (New York: Stein & Day, 1971)

Busschau, W.J., *Measure of Gold* (Johannesburg: Central News Agency, 1949)

Cartright, A.P., *The Gold Miners* (Cape Town: Purnell & Sons, 1962)

Cartright, A.P., *Gold Paved the Way: the Story of the Gold Fields Group of Companies* (London: Macmillan, 1967)

Christman, George, *One Man's Gold: The Letters & Journal of a Forty-Niner* (London: Whittlesey House, 1931)

Cobleigh, Ira U., *Gold, the Dollar and You* (New York: Goldfax, Inc., 1972)

Cousteau, Jacques, *Diving for Sunken Treasure* (Garden City, N.Y.: Doubleday & Co., 1971) (also contains material about silver)

Del Mar, Alexander, *A History of the Precious Metals* (New York: Augustus M. Kelley, 1969)

Einzig, Paul, *The Destiny of Gold* (London: Macmillan, 1972)

Einzig, Paul, *The Future of Gold* (New York: Macmillan, 1935)

Emmons, W. H., *Gold Deposits of the World* (New York: McGraw-Hill, 1937)

Fielder, Mildred, *The Treasure of Homestake Gold* (Aberdeen, South Dakota: North Plains Press, 1970)

Fregnac, Claude, *Jewelry: From the Renaissance to Art Nouveau* (London: Octopus Books, Ltd., 1973)

Friedenberg, Robert, *Gold Coins of the World* (New York: Coin & Currency Institute, 1971)

Go Gold! (New York: Jewelers' Circular-Keystone, 1967)

Gold (New York: Metropolitan Museum of Art, catalog)

Gold 1973 (New York: Consolidated Gold Fields)

Gold Coins (New York: Federal Coin & Currency, 1973)

Green, Timothy, *The World of Gold Today* (New York: Walker & Co., 1973)

Harte, Bret, *Tales of the Gold Rush* (New York: Heritage Press, 1944)

Hobson, Burton, *Historic Gold Coins of the World* (Garden City, N.Y.: Doubleday & Co., 1971)

Hoffman, Arnold, *Free Gold: The Story of Canadian Mining* (New York: Associated Book Service, 1958)

Hogg, Gary, *Lust for Gold* (New York: A.S. Barnes, 1962)

Hoppe, Donald, *How to Invest in Gold Coins* (New Rochelle, N.Y.: Arlington House, 1971)

Hoppe, Donald, *How to Invest in Gold Stocks—and Avoid the Pitfalls* (New Rochelle, N.Y.: Arlington House, 1972)

Hyams, Edward & Ordish, George, *The Last of the Incas* (London: Longmans Green, 1963)

Jennings, Gary, *The Treasure of the Superstition Mountains* (New York: W.W. Norton, 1973)

The Life of Benvenuto Cellini, by himself (London: Phaidon Press, 1949)

The Travels of Marco Polo (New York: Modern Library, 1953)

Moore, Robin and Jennings, Howard, *The Treasure Hunter* (Engelwood Cliffs, N.J.: Prentice-Hall, 1974)

Morrell, W.P., *The Gold Rushes* (Chester Springs, Pa.: Dufour, 1968)

Murray, A.E., *Murray's Guide to the Gold Digging* (Sydney: D.S. Ford, 1956)

Parker, Watson, *Gold in the Black Hills* (Norman, Oklahoma: University of Oklahoma Press, 1966)

Paul, Rodman W., *California Gold: the Beginning of Mining in the Far West* (Lincoln, Nebraska: the University of Nebraska Press, 1947)

Pick, Franz, *Gold—How and Where to Buy and Hold It* (New York: Pick Publishing, 1973)

Rist, Charles, *The Triumph of Gold* (New York: Greenwood, 1969)

Rosenthal, Eric, *Gold! Gold! Gold!* (New York: Macmillan, 1970)

Sédillot, Rene, *Histoire de l'Or* (France: Librairie Arthene Fayard, 1972)

Sutherland, C.H.V., *Gold: Its Beauty, Power and Allure* (London: Thames and Hudson, 1969)

Turner, W.W., *Gold Coins for Financial Survival* (Nashville, Tennessee: Hermitage Press, 1971)

Van Tassell, C. Roger, *The Commercial Demand for Gold in the United States* (Wooster, Massachusetts: Clark University, 1973)

Wharton, David B., *The Alaska Gold Rush* (Bloomington, Indiana: Indiana University Press, 1972)

Windeler, Adolphus, *The California Gold Rush Diary of a German Sailor* (Berkeley, California: Howell-North Books, 1969)

Wise, Edmund, *Gold: Recovery, Properties, Applications* (Princeton, N.J.: D. Van Nostrand, 1963)

Young, Otis E. Jr., *Western Mining* (Norman Oklahoma: University of Oklohoma Press, 1970) (also contains material about silver)

Silver

Banister, D'Arcy and Knostman, Richard W., *Silver in the United States* (Washington, D.C.: U.S. Department of the Interior, 1969)

Butts, Allison and Coxe, Charles D., Ed., *Silver: Economics, Metallurgy and Use* (Princeton, N.J.: D. Van Nostrand, 1967)

De Quille, Dan, *The Big Bonanza* (New York: Alfred A. Knopf, 1947)

Dunning, Charles H., and Peplow, Edward H. Jr., *Silver: From Spanish Missions to Space Age Missiles* (Pasadena, California: Hicks Publishing, 1966)

Finlay, Walter L., *Silver-Bearing Copper* (New York: Corinthian Editions, 1968)

Groseclose, Elgin, *Silver as Money: the Monetary Services of Silver* (Washington, D.C.: Institute for Monetary Research, 1965)

Groseclose, Williams & Associates, *Silver and the Coinage Crisis* (New York: Cyrus J. Lawrence & Sons, 1964)

Leaven, Dickson H., *Silver Money* (Bloomington, Indiana: Principia Press, 1939)

Lewis, Oscar, *Silver Kings (New York: Alfred A. Knopf, 1967)*

Lord, Eliot, Comstock Mining & Miners (Berkeley, California: Howell-North Books, 1959)

Lyman, George, *Saga of the Comstock Lode* (New York: Scribners, 1934)

Magnusen, Richard G., *Coeur d'Alene Diary* (Portland, Oregon: Metropolitan Press, 1968)

Modern Silver Coinage, 1969-1972 (Washington, D.C.: Silver Institute, 1973)

Myers, C.V., *Silver* (La Jolla, California: La Jolla Rancho Press, 1969)

New Silver Technology (Washington, D.C.: Silver Institute, 1974)

Pick, Franz, *Silver—How and Where to Buy and Hold It* (New York: Pick Publishing, 1974)

Preston, Robert, *Building Your Fortune with Silver* (Salt Lake City, Utah: Hawkes Publications, 1973)

Rickenbacker, William F., *Wooden Nickels* (New Rochelle, N.Y.: Arlington House, 1966)

Silver Situation Report (New York: Merrill Lynch, Pierce, Fenner & Smith, Inc., 1974)

Smith, Jerome F., *Silver Profits in the Seventies* (Zurich: Economic Research Corporation, 1971)

A Statistical Analysis of the U.S. Silver Market (New York: Chender Associates, Inc., 1972)

Twain, Mark, *Roughing It* (Garden City, N.Y.: Doubleday, 1966)

Waldorf, John Taylor, *A Kid on the Comstock* (Palo Alto, California: American West Publishing, 1970)

Why Silver Coinage (Washington, D.C.: Silver Institute)

ARTICLES

Economics

Foundation for Economic Education

Webster, Pelatiah, "Not Worth a Continental"

Freeman

Carson, Clarence B., "The Scourge of Inflation," July 1972

Greaves, Percy L., "From Price Control to Valley Forge: 1777-78," February 1972

Grove, Cecil V., "A Short History of Inflation," March 1962

Hazlitt, Henry, "Inflation: A Tiger by the Tail," February 1970

Hazlitt, Henry, "Uruguay: Welfare State Gone Wild," April 1969

Lynch, Alberto Benegas, "The Argentine Inflation," December 1972

McBain, Hughston M., "A Merchant's Appraisal of Inflation," June 1959

Gold & Silver Newsletter

"How Runaway Inflation Devastated a Nation," February 28, 1973

Wheeler, Timothy J., "An Analysis of Runaway Inflation in America," July 1972

Journal of Commerce

"Will Inflationary Fires Be Fanned?," June 15, 1973

Money Manager

Friedman, Milton, "Who Will Lead U.S. Out of Inflation if Fed Does Not?," March 11, 1974

Love, Roger, "Inflation Rates Worldwide Jump from Trot to Canter, Soon May Be at Gallop," April 1, 1974

Love, Roger, "Inflation Throughout the Industrialized West Already Outdistancing Predictions for Year," March 11, 1974

Nation's Business

"When Inflation Runs Wild," January 1965

New York Times

Bladen, Ashby, "Too Much Credit Built Up Too Fast," March 10, 1974

Butterfield, Fox, "Inflation Breathes New Life into Japan's Pawnshops," January 31, 1974

"Cost of Dinner for 4 Up 17% in Year," August 14, 1973

Dullea, Georgia, "Rising Food Costs Hit Nearly Everyone—but Force Varies," February 14, 1974

Groppelli, A.A., "Inflation Around the World," August 18, 1973

Kaufman, Henry, "Inflation (Not Oil) the Real Culprit," January 27, 1974

"Runaway Prices Jolt 3 Latin Countries," January 26, 1973

"Strikes and Shootings Spread Across Chile," June 22, 1973

Reader's Digest

Andrew, T. Coleman, "Here's How Inflation Has Victimized You," March 1959

Tax Foundation, "The Case of the Vanishing Pay Raise," March 1974

South African Financial Gazette

Lyons, Stephen, "Inflation Hits at Those Burial Costs as Well," February 15, 1974

U.S. News & World Report

Flieger, Howard, "Fright: A Flashback," December 24, 1973

"Seven Decades of Inflation," June 4, 1973

Wall Street Journal

Hartley, William D., "In Indonesia, Wealth Flows In, but Masses Don't Get Much of It," February 26, 1974

Malabre, Alfred L., "The Awful Year Inflation Ran Wild," August 21, 1973

Martin, Everett G., "Wrecking Chile to Build Marxism," November 13, 1972

Martin, Everett G., "The Crucial Year for Chile's Allende," July 6, 1973

Martin, Everett G., "Overthrow of Allende Seen as No Guarantee of a Return to Stability," September 12, 1973

Sorrells, Eugene, "It's Time to Talk Turkey for Thanksgiving: Main Dish to Cost Twice as Much This Year," October 22, 1973

Gold

Allentown Call (Pennsylvania)

Koch, John H., "Rush to Buy Gold, Silver Spreads to Lehigh Valley," March 10, 1974

American Jewelry Manufacturer
Green, Timothy, "Gold: Other Gold Markets, or From Dubai to India on a Dhow," January 1973
American Metal Market
Cohn, J.G., "Precious Metals New Applications Increasing," December 8, 1969
Henderson, Richard Jr., "The Story of Industrial Gold," December 12, 1973
Barron's
Bleiburg, Robert M., "Some 'Barbarous Relic,' " January 24, 1972
Bleiburg, Robert M., "Store of Value: The 'Barbarous Relic' Has Reaffirmed Its Worth," May 13, 1974
Thomas, Dana L., "Solid Gold," November 8, 1971
Business Week
"Gold Fever Spurts a Rush for the Mines," June 9, 1973
Chicago Sun-Times
"It's a Big Rush on Coins," March 2, 1974
Executive
Fellows, Pat, "Yes, the Optimists See Gold at $200-$300 an Ounce," July-August 1973
Financial World
Berton, Lee, "Who Sets the Gold Price?," February 20, 1974
Stinson, Richard J., "What If They Legalize Gold?," February 20, 1974
Freeman
Hazlitt, Henry, "Back to Gold?," October 1965
Gold & Silver Newsletter
Pick, Franz, "The Triumph of Gold," January 1974
Pick, Franz, "War, Inflation and Gold," November 15, 1973
Holt Investment Advisory
Holt, T.J., "Singing in the Rain," February 1, 1974
London Mining Journal
"Gold and the Oil Crisis," February 1, 1974
Long Island Press
Van Hoffman, Nicholas, "Smart Money Out of Paper, Into Coins," March 8, 1974

Metal Progress

Tugwell, Gilbert L., "Industrial Applications for the Noble Metals," June & July 1965

National Geographic

Burg, Amos, "Along the Yukon Trail," September 1953

Kenelt, F.L., "Tutankhamen's Golden Trove," October 1963

Stenuit, Robert & Littlehales, Bates, "Priceless Treasures of the Spanish Armada," June 1969

Wagner, Kip & Imboden, Otis, "Drowned Galleons Off Florida Yield Spanish Gold," January 1965

White, Peter T., "Gold, The Eternal Treasure," January 1974

New York Times

"A New Bonanza for South Africa: Soaring Gold Prices," November 25, 1973

Borders, William, "Gold Rush Beginning at Canadian Banks," March 2, 1974

Farnsworth, Clyde H., "Small Investor Turns to Gold," March 28, 1974

Farnsworth, Clyde H., "Strike on Bourse Moves Gold Trading Into Street," April 15, 1974

"Gold Frenzy," February 23, 1974

Hawthorne, Peter, "Mining Is Stressed in South Africa," January 27, 1974

Maidenberg, H.J., "The World of Gold," March 10, 1974

Playboy

Pick, Franz, "Gold," August 1969

Reader's Digest

Hauser, Ernest O., "Gold: King of Metals," June 1973

Seminars on Drug Treatment

Bland, John H., "Drug Treatment of Rheumatoid Arthritis," September 1971

Toronto Globe & Mail

Needham, Richard J., "Farewell to the Yankee Dollar," July 17, 1973

Wall Street Journal

"Gold Gains in Lustre for Investors, Muddies the Monetary Waters," July 23, 1973

Metz, Tim, "For a Prospector, Gold Is Among the Minerals Most Difficult to Find," July 27, 1973

Vicker, Ray, "High Gold Price Aids South Africa's Mines, But Production Drops," July 24, 1973

Wall Street Transcript

"Gold: A Roundtable Discussion," December 31, 1973

Washington Post

Sear, John, "Gold (Coin) Rush Flourishing in DC," February 24, 1974

Touhy, William, "Rich, Poor Joining the Gold Rush in Geneva," March 31, 1974

Windsor Star (Canada)

Rolland, Keith, "You may Not Be Aware . . Windsor Has a Gold Rush," February 28, 1974

Silver

Barron's

Thomas, Dana L., "Three Bags Full: Silver Coins Have Shown Handsome Gains, Too," January 21, 1974

Business Week

"Silver Coins: A Sterling Idea for Investors," June 2, 1973

Denver Post

Wilkinson, Bruce, "Investors Climb Onto Silver-Coin Bandwagon," October 5, 1973

Engineering & Mining Journal

Lindstrom, Philip M., "Silver—For 1973, Prices Will Stay Volatile on Rising Trend," March 1973

Lindstrom, Philip M., "Silver—Speculation Adds Spice to Volatile Market," March 1974

Gold & Silver Newsletter

"Silver Shortage Worsens: Consumption-Production Gap Widens," April 30, 1973

"U.S. Silver Consumption Up 11.1%," January 31, 1973

Houston Post

Miller, Dean, "Silver Makes a Mint," August 5, 1973

National Observer

Paulson, Morton, "Investors Look for the Silver Lining," July 21, 1973

New York Times

Walker, Robert, "Treasury Sells the Last of Its Silver," November 11, 1970

Pacific Coast Coin Exchange

Carabini, Louis M., "The Case for Silver" (1974)

Silver Institute Letter

"Booming Trade in Silver Collectibles and Gifts," November 1973
"Canada's Silver Olympic Coins Due Soon," October 1973
"Dupont Introduces New Silver X-Ray Film," December 1973
"End of a Hurricane," October 1971
"Flexible Silver Electrode Stretches," November 1973
"Japan's Silver Consumption Rises Again; Production Down," February 1973
"More Silver Being Used in Coinage," June 1973
"More Silver Needed by New Strategic Defense Systems," April 1973
"New Electronics Technology Uses Silver." July-August 1973
"New Uses for Silver," January 1972
"Non-Phosphate Detergents from Silver," September 1973
"The Power of Silver Catalysts," May 1973
"Silver Defrosts Auto Rear Window," September 1972
"Silver Developments in Russia," June 1972
"Silver in First-Aid for Burns," November 1973
"Silver in Nuclear Power Plants," December 1971
"Silver in Solar Power," February 1972
"Silver in Space," September 1971
"Silver Plays Medical Role," February 1973
"Silver Purifies Water for Drinking," May 1972

Silver Market Letter

Demoraes, Albert, "The Silver Rush," March 21, 1974
"Hunt to Corner Silver?," March 7, 1974

Wall Street Journal
"Surge in Silver Demand Pushes the Metal's Price to $6.77 an Ounce at London and $6.70 in the U.S.," February 27, 1974

Washington Star News
Oppenheimer, Jerry, "Not Just for the Wealthy," April 15, 1974

PUBLICATIONS

Barrons (weekly; $21 a year)
22 Cortlandt Street
New York, N.Y. 10007

Bulletin (quarterly; no charge)
Swiss Credit Bank
Box 8021
Zurich, Switzerland

Coin World (weekly; $7.50 a year)
Box 150
Sidney, Ohio 45365

Consumer/Wholesale Price Index (monthly; $16.50 a year)
U.S. Department of Labor
Bureau of Labor Statistics
Washington, D.C. 20210

Dow Theory Letters (36 times a year; $95 a year)
Box 1759
La Jolla, California 92037

ERC World Market Perspective (9-12 times a year; $96 a year)
Box 91491
West Vancouver, B.C.
Canada

Freeman (monthly; no charge)
Foundation for Economic Education
Irvington-on-Hudson, New York

Gold Bulletin (quarterly; no charge)
Chamber of Mines
5 Hollard Street
Johannesburg, South Africa

Gold & Silver Newsletter (monthly; $36 a year)
Pacific Coast Coin Exchange
3711 Long Beach Boulevard
Long Beach, California 90807

Green's Commodity Market Comments (24 times a year; $125 a year)
Box 174
Princeton, New Jersey 08540

Holt Investment Advisory (twice a month; $144 a year)
277 Park Avenue
New York, N.Y. 10017

International Currency Review (6 times a year; $75 a year)
11 Regency Place
London SW1P 2EA
England

International Harry Schultz Letter (every 3 weeks; $200 a year)
Box 161
Basel 4002
Switzerland

Journal of Commerce (weekdays; $52 a year)
99 Wall Street
New York, N.Y. 10005

London Mining Journal (weekly; $34 a year)
15 Wilson Street
Moorgate, London E.C. 2
England

Mineral Industry Surveys (for gold & silver—monthly; no charge)
U.S. Bureau of Mines
Washington, D.C. 20240

Monetary Trends (monthly; no charge)
Federal Reserve Bank of St. Louis
Box 442
St. Louis, Missouri 63166

Money Manager (weekly; $96 a year)
77 Water Street
New York, N.Y. 10005

Myers Finance & Energy (twice a month; $135 a year)
Box 5531 Station A
Calgary, Alberta Canada T2H 1X9

National Economic Trends (monthly; no charge)
Federal Reserve Bank of St. Louis
Box 442
St. Louis, Missouri 63166

New York Times (daily; $117.60 a year)
229 West 43rd Street
New York, N.Y. 10036

Northern Miner (weekly; $15 a year)
77 River Street
Toronto, Canada M5A 3P2

Pick's Currency Yearbook (annual; $100)
21 West Street
New York, N.Y. 10006

Pick's World Currency Report (monthly; $300 a year)
21 West Street
New York, N.Y. 10006

Silver Institute Letter (monthly; no charge)
Suite 1138
1001 Connecticut Avenue
Washington, D.C. 20036

South African Financial Gazette (weekly; $90 a year)
8 Empire Road Extension
Auckland Park
Johannesburg, South Africa

Viewpoints (occasional; $5 a year)
Institute for Monetary Research
1010 Vermont Avenue, N.W.
Washington, D.C. 20005

Wall Street Journal (weekdays; $35 a year)
22 Cortlandt Street
New York, N.Y. 10007